# Residential Windows

## *A Guide to New Technologies and Energy Performance*

John Carmody

Stephen Selkowitz

Lisa Heschong

W.W. Norton & Company
New York • London

A NORTON PROFESSIONAL BOOK

The text of this book is composed in Walbaum,
with the display set in Adobe Berthold Walbaum, Avant Garde, and Tekton
Manufacturing by Quebecor/Kingsport
Book design by John Carmody
Cover illustration: Marvin Windows & Doors

Library of Congress Cataloging-in-Publication Data

Carmody, John
        Residential windows : a guide to new technologies and energy
    performance / John Carmody, Stephen Selkowitz & Lisa Heschong.
            p.     cm.
    "A Norton Professional Book."
    Includes bibliographical references and index.
    ISBN 0-393-73004-2 (pbk.)
    1. Windows.  2. Dwellings—Insulation.  I. Selkowitz, Stephen.
II. Heschong, Lisa.  III. Title.
TH2275.C37  1996
690' . 1823—dc20                                        96-28781
                                                        CIP

W. W. Norton & Company, Inc., 500 Fifth Avenue, New York NY 10110
    nttp://www.wwnorton.com
W. W. Norton & Company Ltd., 10 Coptlc Street, London WC1A 1PU

0 9 8 7 6 5 4 3

# Contents

Acknowledgments   4

A Guide to Using This Book   5

1. Introduction to Windows in Residential Buildings   9

2. Energy Performance Characteristics of Windows   24

3. Window Glazing Materials   41

4. The Complete Window Assembly   87

5. Design Implications with Energy-Efficient Windows   110

6. Window Selection Considerations   143

Appendix A: Energy Performance Calculation Assumptions   180

Appendix B: Overview of the NFRC Program   185

Appendix C: Specifying Product Performance   188

Appendix D: Resources   190

Appendix E: Codes and Standards   193

Glossary   195

References   202

Index   210

# Acknowledgments

This book was developed with support from the U.S. Department of Energy's Windows and Glazing Research Program within the Office of Building Technology, Community and State Programs. This support was provided through Lawrence Berkeley National Laboratory and Oak Ridge National Laboratory. In partnership with the building industry, the U.S. Department of Energy supports a range of research, development, and demonstration programs, as well as education and market transformation projects designed to accelerate the introduction and use of new energy-saving window technologies, including many of those described in this book.

Many people contributed their time and talent toward the completion of this project. The book would not have been possible without the continued support and encouragement of Sam Taylor, manager of the Windows and Glazing Research Program at the U.S. Department of Energy. Sam has overseen the project through its many phases and helped to shape the final product.

Much of the technical material in the book was drawn from window and glazing research completed over the years at Lawrence Berkeley National Laboratory (LBNL). Robert Sullivan managed the first phases of the project and later contributed to the energy performance analysis portion of the book. Dariush Arasteh, Brent Griffith, and others at LBNL made important technical contributions.

A very important contribution was made to the project by the National Fenestration Rating Council's annual energy performance task group led by James Larsen and Brian Crooks of Cardinal/IG. Both Brian and Jim helped to develop the NFRC rating method, and Brian spent countless hours providing additional computer simulations and technical assistance for the book. Much of the energy performance information in the book is based on Brian's work.

Many people reviewed and contributed suggestions to the project. A few, however, spent considerable time providing very careful technical reviews and contributing necessary material. We are grateful to Ross McCluney at the Florida Solar Energy Center; John Hogan of the Department of Construction and Land Use, City of Seattle; and Carl Wagus and Richard Walker of the American Architectural Manufacturers Association. Dragan Curcija of the University of Massachusetts assisted with the references. During earlier stages of the project useful reviews were received by G. Z. Brown at the University of Oregon, Chris Gueymard at the Florida Solar Energy Center, and Chris Mathis, of Mathis Consulting.

We appreciate the many comments and suggestions we received from people associated with the National Fenestration Rating Council, including John Rivera, Susan Douglas, and William duPont. In addition, many members of the window industry contributed review comments and materials for the book. They include: Michael Koenig, Andersen Windows; Michael Curtis, Cardinal IG; Erik Ekstrom, National Wood Window and Door Association; James Krahn, Marvin Windows & Doors; David Duly, Libby Owens Ford; James Benney, Primary Glass Manufacturing Council; Herb Yudenfriend, Suntrol Products; and Ron Saunders, Ultra-Glass Products. We also thank Debra Brunold of Jeld-Wen, Mike Massinople of Velux, Dave Weiss of Andersen Windows, and Nicole Welu of Marvin Windows & Doors for their assistance in obtaining photographs.

The illustrations and design for the book were done by John Carmody. This book has benefited greatly from the suggestions and editing of Nancy Green, our editor at W.W. Norton.

John Carmody
Stephen Selkowitz
Lisa Heschong

# A Guide to Using This Book

This book is about windows in homes. Everyone is familiar with how windows provide views, fresh air, and light, and they recognize the important role of windows in the aesthetic appearance of buildings. What is less obvious is that windows have undergone remarkable technological changes in the last fifteen years. Some of these new technologies are so effective in reducing the entry of hot summer sun and preventing the loss of warmth in winter that windows must be looked at in a completely new way. However, consumers are faced with a range of technically complex choices, and designers often do not have the tools to evaluate performance or understand how new products might change assumptions about design.

Windows are normally selected based on four groups of considerations—appearance, function, energy performance, and cost. Information to make these judgments comes mainly from product literature as well as from architects, builders, and salespeople. Because there has been confusion regarding the true benefits and impact on energy performance, there was a need to compare products based on standardized procedures that are fair and technically credible. To address this problem the National Fenestration Rating Council (NFRC) was formed in 1989. The NFRC develops methods of rating and labeling windows, and then certifies products for a number of thermal and optical properties that affect energy use, comfort, and other factors. The U.S. Department of Energy has contributed scientific and technical expertise toward the development of the NFRC procedures.

The purpose of this book is to assist consumers, designers, and builders in understanding the new window products and their energy performance implications. Our hope is that this understanding will lead to greater use of these new products with benefits to the owners and to society. The broader audience for the book includes anyone who needs to be informed about windows—regulators, standards developers, utilities, and the researchers, manufacturers, and suppliers in the window industry itself.

The book introduces the window technologies, explores the implications of these new technologies on residential design, and then provides a means for selecting appropriate windows. The introductory chapter provides an overview. Chapter 2 presents the basic energy performance character-

## Window selection process

**Gather information on window products**
- Product literature
- Store visit
- Product labels (NFRC)
- Advice from experts (architects, builders, salespeople)

**Understand concepts and new technologies**
- See Chapters 1–4 of this guidebook

**Consider how new window technologies influence design**
- See Chapter 5 of this guidebook

**Evaluate options based on your selection criteria**
- See Chapter 6 of this guidebook
- Use NFRC ratings and, if needed, computer programs to determine energy performance

**Make selection and specify products**
- See Appendix C of this guidebook for specs

Some improved technologies can lead to changes in basic house design assumptions.

Improved window technologies
- High insulating value
- Low solar heat gain
- Low air leakage

Improved performance
- Heating energy savings
- Cooling energy savings
- Greater comfort
- Less condensation

Design implications
- Amount of glazing is less critical
- Window orientation can be less important
- Thermal shades and shutters are not necessary
- Shading devices have less impact

istics of windows and describes how these are determined. Chapter 3 describes glazing materials and new technologies in detail. Chapter 4 addresses the complete window assembly, which includes window operation, frame materials, and installation.

Chapter 5 reviews traditional window design issues and explores the new design implications of using high-performance windows. When improved building technologies are introduced, often they are simply applied to the same architectural design approaches that were used in the past. When the technological changes are significant enough, they can sometimes make obsolete some of the traditional patterns and assumptions about design. Energy-efficient windows are one such technology, and this book attempts not only to describe improvements but to indicate their implications for how we think about house design.

Finally, Chapter 6 summarizes the entire range of window selection considerations and provides a checklist for use by designers, builders, and homeowners. Even though these groups have access to much information about new window technologies, it is difficult for them to determine how particular windows will perform on a specific house in a specific climate. For this reason, methods for assessing annual energy performance are described in Chapter 6.

## Understanding Window Features

Throughout the book we refer to a series of prototypical window designs. These are meant to be "generic" windows that are illustrative of a broad range of older products that you will find in existing homes, currently available window products sold today, and a few emerging technologies that are not yet commonplace. We describe the products both by some of their key features (e.g., low-E coating, E=.08) as well as by thermal parameters (e.g., U = .33 Btu/hr-sq ft-°F).

As you shop for windows, you will discover that there will often be a wide range of different features or combinations of features that can achieve the same thermal properties. You will also see that products with similar descriptions may have very different thermal properties because one or two seemingly small and "invisible" (but important) changes have been made (e.g., changing the air space of double glazing from 1/2- to 1/4-inch and omitting argon gas fill). We provide enough technical detail for the committed reader to understand and address most of these subtleties, but for those who do not wish to do so, relying on the NFRC labeled thermal properties should be adequate.

NFRC certified *whole window* numbers are used throughout this book. However, when you look for windows on the market, you will no doubt be confronted with a variety of other "facts" and figures. Whenever possible, ask to see numbers on the NFRC labels so that comparisons can be made on an "apples-to-apples" basis. In some sections of this book we provide thermal properties for glazings alone, which may also be useful in sorting out competing claims in the marketplace. Throughout the book we refer to selected combinations of twelve prototypical windows; for the benefit of the more technical readers, these are described in more detail in Appendix A. As previously mentioned, these windows are intended only to be illustrative of a wide range of commercially available products.

## Determining Energy Savings

We illustrate the energy and cost consequences of various window technology and design decisions by using results from computer simulations of a prototypical house design in three U.S. climates: Phoenix, Arizona (cooling dominated), St. Louis, Missouri (heating and cooling), and Madison, Wisconsin (heating dominated). The various thermal properties and operational characteristics of the prototypical house are described in detail in Appendix A. Although your house will no doubt be different from this typical house, extensive studies have shown that the comparative results which show the relative performance of any two windows on the standard house will almost always provide useful guidance for your house as well. However, the text accompanying the figures provides some warnings and cautions with respect to use of the information.

Many readers will want to go beyond comparative guidance and determine, for example, the payback period for new high-efficiency windows. To more accurately determine the actual performance of different specific windows in your house, some type of calculation method must be used that accounts for the climate, the surroundings, the house design and operating parameters, heating and cooling system, utility costs, etc., as well as the properties of the windows being compared. In the past such a calculation would be time consuming, would require pages of tables and explanation, and would still be approximate at best. The widespread availability of affordable computers has changed this situation and new, more user friendly software allows architects, builders, and consultants to perform these calculations. In addition, recently some of these tools have become available via the Internet on the World Wide Web. It is no longer necessary to install and maintain software to do the calculations—they are available with any web browser on any computer operating system. We have chosen to direct those readers wanting detailed calculations of window energy performance to a simulation tool, RESFEN, which is available via the Web, to obtain an on-line calculation. See Appendix D for more information about access to RESFEN on the Web.

Many window manufacturers are beginning to place their product information on the Web. Use one of the search engines commonly available in the browsers to find out if the information you want is available. As of the publication date of this book, the Web is evolving and growing rapidly, and access to the Web is becoming more widespread. If you do not have a computer or Internet access, consider asking a friend or looking for public access from a local library or school.

New calculation tools can help you to estimate the energy savings potential of better windows. If the resulting payback on your investment is short, then it clearly makes sense to purchase the windows. But even if the conclusion is that these windows are not immediate cost savers, there are many other good reasons to proceed with the investment. (See Chapter 6 for further discussion of cost issues.)

## Finding the Elusive "Right Answer"

This book is intended to help you make better, more informed decisions that will be important to your future comfort and to your finances. However, there are so many complex issues that are difficult to balance in selecting the "best" window that you are likely to find there is no absolutely "correct" answer. How much more are you willing to pay for appearance and comfort? The selection of new (or retrofit) windows is a big investment for most homeowners and one that deserves critical and informed attention. Ultimately, decisions must be made based on many trade-offs. This book is intended to help you understand the options and their consequences and to make the best decision within the scope of those trade-offs and constraints.

# CHAPTER 1

# Introduction to Windows in Residential Buildings

Windows are possibly the most complex and interesting elements in residential design. They provide light and fresh air, and offer views that connect the interior spaces with the outdoors. However, windows have also represented a major source of unwanted heat gain in summer and significant heat loss in winter. Today, remarkable new window products and technologies have changed the energy performance of windows in a radical way. Of course, this has a powerful impact on how we select windows, but it also affects the way we use windows in house design.

Figure 1-1. New window technologies improve energy efficiency and influence home design. (Photo: Andersen Windows.)

Until about the end of World War II, housing in the United States was designed with an understanding of site and climate. Although the windows were not particularly energy efficient, traditional house designs evolved that took advantage of the natural elements of sunlight, wind, the earth, and vegetation to help provide light, heating, cooling, and ventilation. A house built in Florida looked quite different from a house built in Maine, reflecting their climate differences. While these buildings were not always comfortable by today's standards, energy use was minimized as much as possible. During this period, many elements of exterior and interior design evolved, in part because of the need to either shade windows from the summer sun or to protect from cold drafts and high heat loss in winter.

From 1950 to 1970, the availability of very inexpensive energy combined with the use of powerful mechanical heating and cooling systems led to the construction of homes that were not climate sensitive and required a large amount of energy to maintain comfort. Most homes were not designed to take advantage of sunlight and natural ventilation, and windows remained inefficient during this period.

In the 1970s, rapid increases in energy costs occurred, combined with more concern about the environmental impacts of building design and operation. This led to a resurgence of interest in the traditional patterns of designing with climate and site. During this period, design approaches in colder

Figure 1-2. Skylights and roof windows can provide daylight but limit undesirable heat gain. (Photo: Velux-America Inc.)

Figure 1-3. More efficient windows provide greater thermal comfort. (Photo: Andersen Windows.)

Figure 1-4. With high-performance glazings, the benefits of large window areas can be obtained without the energy penalty associated with older, less efficient windows. (Photo: Marvin Windows & Doors.)

regions included reshaping building layout and orienting windows to capture the maximum amount of sunlight in every room, various schemes to store and distribute solar heat, and movable insulation over windows to keep heat from escaping at night. These approaches reflected the fact that windows were necessary for light and to capture solar heat, but they came with the drawback of significant heat loss.

These passive solar and self-sufficient house designs never became mainstream practice. However, the concern over energy and environment since 1970 has led to significant improvements in house performance based on the use of more efficient building envelopes and mechanical systems. During this period, windows have undergone a technological revolution. They are no longer the weak link in energy-efficient home design. As Figure 1-5 illustrates, if windows are highly efficient, there is no significant energy penalty when total glazing area is increased. In fact, for the most efficient window in the chart, the winter heating load is reduced as the glazing area is increased. As will be demonstrated throughout the book, there are similar significant improvements when higher-performance glazings are used in cooling-dominated climates.

It is now possible to have expansive views and daylight without sacrificing comfort or energy efficiency. This remarkable change has two important effects. First, any house can be made considerably more energy efficient by using high-performance windows. Second, and possibly more important,

Figure 1-5. Impact of window glazing area on annual heating season energy use in Madison, Wisconsin.

Case 1: Glazing area is 5% of floor area (77 sq ft).

Case 2: Glazing area is 15% of floor area (231 sq ft).

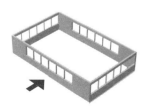

Case 3: Glazing area is 25% of floor area (385 sq ft).

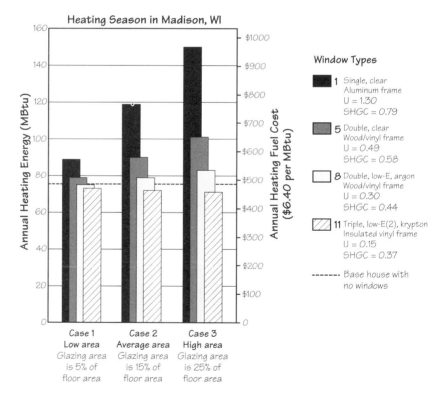

Note: The annual energy performance figures shown here are for a typical 1540 sq ft house. U-factor and SHGC are for total window including frame. House and windows are described in Appendix A. MBtu=millions of Btu.

technologically advanced windows perform so much better and differently than their predecessors of just ten years ago that many of the assumptions of both traditional and more recent energy-efficient design must be reexamined. Now that the window unit itself has substantial insulating value and still provides solar heat gain in winter, its net annual energy use can be better than a well-insulated wall in some cases.

## A SHORT HISTORY OF WINDOWS

In a way, windows are a luxury. Primitive homes were often built without windows. A door, of course, is essential to let people and contents in and out. But windows are a refinement, an amenity, to make the place more livable. As such, they have been under continual development throughout the ages.

A smoke hole might be considered the earliest form of a window. It let the smoke from cooking and heating fires escape out the wall or roof, greatly enhancing indoor air quality. Inadvertently, the hole also provided a shaft of daylight that brightened the general gloom of the interior, and, of course, allowed most of the heat to escape.

A single shuttered opening came next. It was like another door—a hole in the wall with an opaque cover that could be opened to let in light and air, along with intruders, rain, insects, and dust. It could be closed for security, darkness, and protection from the elements.

Figure 1-6. Limitations in the size of glass resulted in early architectural innovations to provide larger views.

### A Room with a View

The addition of translucent materials, such as oiled paper or an animal skin, framed into the window hole offered more control options. The shutter itself might be made of translucent material, thereby creating an operable window. Not until the advent of transparent window glass, which was first used in Roman times, could windows provide daylight, wind control, and view, all at the same time.

The largest known piece of Roman glass was three feet by four feet (0.9 by 1.2 m), installed in a public bath in Pompeii. By the Middle Ages, Venice had become established as the premier center of glass making, for both decorative glassware and clear window glass. Small panes of flat glass could be produced by first blowing a bubble or cylinder of glass, cutting it open while still hot, and then rolling it out flat. This technol-

13

ogy was brought to the New World and used to produce most of the glass for colonial American windows. These small panes of glass, pieced together into multiple frames, have become one of our enduring domestic images.

## The French Glass Revolution

A new technique to cast plate glass was developed in France in the 1600s. The finer quality and larger sizes of glass that became available with this process greatly popularized the use of glass, both for grand mirrors, such as those at the Hall of Mirrors at Versailles Palace, and for large windows, as epitomized by the "French door."

Many innovations in the production of glass were seen in the nineteenth century, making larger, stronger, and higher-quality glass ever more available to the general public. While the size and number of windows in buildings increased

Figure 1-7. Elaborate architectural elements have developed in order to control light, air flow, sound, and comfort near windows. (Photo: John S. Reynolds, AIA.)

dramatically, there was still essentially one type of glass available: clear, single-pane glass.

The clear, single-pane glass stopped the wind and allowed light to enter, but all of the other subtleties in window control were provided by additional devices such as overhangs, trellises, awnings, shutters, storm windows, security grates, insect screens, venetian blinds or roll-down shades, light curtains, or heavy drapes. A window became an elaborate and decorative architectural system for controlling all the physical and emotive forces that converged at that hole in the wall. All of these controls have been incorporated into the general aesthetics of windows.

## Modern Developments

In the 1950s the technique of producing float glass was developed (molten glass "floats" over a tank of molten tin), which provides extremely flat surfaces, uniform thicknesses, and few if any visual distortions. This float glass is commonly used in most residential windows today. This was a key breakthrough that has become important decades later because the high-quality surface of float glass is required for the application of thin coatings that are commonly used in windows today.

Before 1965, single-glazed windows with storm windows and screens were prevalent in the United States. The most important trend in windows between 1965 and 1980 was a significant change to insulating glazing (two panes of glass sealed together with an air space in between). Although double-glazed units were developed before 1965, the seals were not maintained consistently, resulting in some product failure. In the late 1970s, in response to the energy crisis of that time, triple-glazed units were developed as well. The change to insulating glazing was accelerating by 1980—the market reflected this trend with window sales of 50 percent single glazing, 45 percent double glazing, and 5 percent triple glazing. Today, nearly 90 percent of all residential windows sold are insulated glazing (two or more layers). The market for conventional triple-glazed windows has diminished because a number of other technological breakthroughs have created a new series of energy-efficient windows with better performance and fewer drawbacks (although triple glazing is used in combination with other features).

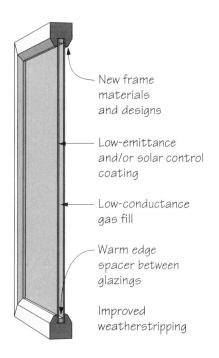

New frame
materials
and designs

Low-emittance
and/or solar control
coating

Low-conductance
gas fill

Warm edge
spacer between
glazings

Improved
weatherstripping

Figure 1-8: Technological advances have significantly improved window energy performance in recent years.

# TECHNOLOGICAL IMPROVEMENTS TODAY

A progression of innovations has integrated more elements of control into the window assembly or the glass itself. Some technological innovations that are appearing in today's fenestration products are described briefly below (see Chapters 3 and 4 for more detail on these technologies).

- **Glazing unit structure**

  Multiple layers of glass or plastic films improve thermal resistance and reduce the heat loss attributed to convection between window layers. Additional layers also provide more surfaces for low-E or solar control coatings.

- **Low-emittance coatings**

  Low-emittance or low-E coatings are highly transparent and virtually invisible, but have a high reflectance (low emittance) to long-wavelength infrared radiation. This reduces long-wavelength radiative heat transfer between glazing layers by a factor of 5 to 10, thereby reducing total heat transfer between two glazing layers. Low-emittance coatings may be applied directly to glass surfaces, or to thin sheets of plastic (films) which are suspended in the air cavity between the interior and exterior glazing layers.

- **Low-conductance gas fills**

  With the use of a low-emittance coating, heat transfer across a gap is dominated by conduction and natural convection. While air is a relatively good insulator, there are other gases (such as argon, krypton, and carbon dioxide) with lower thermal conductivities. Using one of these nontoxic gases in an insulating glass unit can reduce heat transfer between the glazing layers.

- **Warm edge spacers**

  Heat transfer through the metal spacers that are used to separate glazing layers can increase heat loss and cause condensation to form at the edge of the window. "Warm edge" spacers use new materials and better design to reduce this effect.

- **Thermally improved sash and frame**

  Traditional sash and frame designs contribute to heat loss and can represent a large fraction of the total loss when high-performance glass is used. New materials and improved designs can reduce this loss.

- **Solar control glazings and coatings**

  To reduce cooling loads, new types of tinted glass and new coatings can be specified that reduce the impact of the sun's heat without sacrificing view. Spectrally selective glazings and coatings absorb and reflect the infrared portion of sunlight while transmitting visible daylight, thus reducing solar heat gain coefficients and the resulting cooling loads. These solar control coatings can also have low-emittance characteristics.

- **Improved weatherstripping**

  Better weatherstrips are now available to reduce air leakage, and most are of more durable materials that will provide improved performance over a longer time period.

## Overall Impact of Improving Window Efficiency

Everyone has a short-term, local reason to be concerned about the annual energy performance of windows, as well as a longer-term, national and global reason. In the short term, we pay the utility bills—each dollar wasted on inefficient building performance could be better spent elsewhere. At the national level, our collective decisions and behavior regarding energy use influence our nation's energy security, global prices for energy, as well as pollution emissions and contributions to global warming.

The impact of changing to energy-efficient windows can be significant even though windows usually comprise a small percentage of the total building envelope by area. Figures 1-9 and 1-10 show the heating and cooling season energy use in a typical house in several U.S. climates. In each climate, the annual energy use with four types of windows is illustrated.

Such energy efficiency improvements can have a large impact on overall national energy consumption. There are nineteen billion square feet of windows in our nation's homes—

stacked end to end on top of each other these windows would form a tower one million miles high! These windows create additional energy bills for our nation's homeowners of $9.3 billion per year. Each time a window is specified for a new or existing house, the decisionmaker has the opportunity not only to save some money on the monthly energy bills but to shrink this national energy cost as well. The cumulative effect of individual decisions can be significant. Studies at Lawrence Berkeley National Laboratory suggest that if all windows purchased over the next fifteen years incorporated low-E coatings, gas fills, and a few other readily available efficiency improvements, our collective annual energy bill could be reduced by 25 percent or over $2 billion per year by 2010.

These potential savings are so significant that it is useful to think of them as a new source of energy supply. It is increas-

Figure 1-9. Annual heating energy performance with different windows in four U.S. climates.

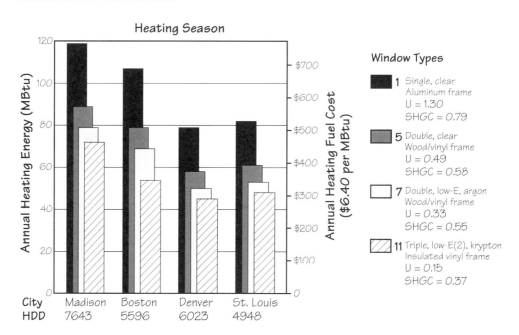

Note: The annual energy performance figures shown here are for a typical 1540 sq ft house. U-factor and SHGC are for total window including frame. House and windows are described in Appendix A. MBtu=millions of Btu. HDD=heating degree days.

ingly costly and environmentally disruptive to provide new energy supplies for a growing economy. All large-scale non-renewable energy sources create significant short-term and long-term environmental impacts. Every ton of coal, cubic foot of gas, or barrel of oil consumed releases carbon dioxide into the atmosphere and contributes to global warming. Consider the following two alternatives to create the equivalent of 36 million barrels of oil:

1. An off-shore oil platform contains ten operating wells each producing 10,000 barrels of oil per day. These are costly capital-intensive operations with the risk of coastal oil spills and have a nominal life of ten years before the fields are pumped dry. Over its ten-year useful lifetime this platform will provide approximately 36 million barrels of oil.

Figure 1-10. Annual cooling energy performance with different windows in four U.S. climates.

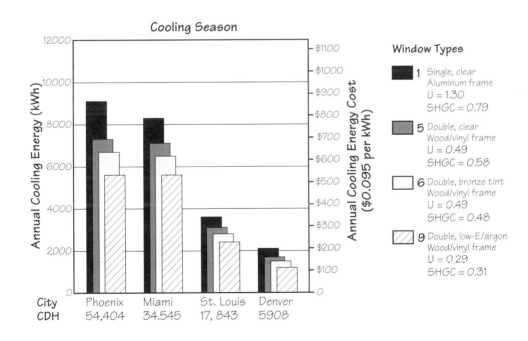

Note: The annual energy performance figures shown here are for a typical 1540 sq ft house. U-factor and SHGC are for total window including frame. House and windows are described in Appendix A. kWh=kilowatt hours. CDH=cooling degree hours.

2. A low-E coating plant provides coated glass to a window manufacturer. The coating plant can produce 20 million square feet of coated glass per year and has a nominal life of ten years. The windows produced with the low-E glass save energy in the home, compared to uncoated glass. The cumulative energy savings from these low-E windows over a twenty-year period is also the equivalent of 36 million barrels of oil!

The energy efficiency features that have already been incorporated into residential construction have greatly reduced the rate of growth of total residential energy consumption. This record can be improved upon significantly by greater use of technologies now on the market such as high-performance windows.

## THE FUTURE

It is clear that there has been great progress in improving window energy efficiency in recent years, but even greater potential lies ahead. In the 1970s, special coatings and gas fills seemed a long way off. Now they are standard products. Research and development time for advanced window technologies has been shortening, and their introduction into the commercial market has been accelerating.

It is difficult to predict which of the many research directions now being pursued will find a secure place in the market. Researchers continue to experiment with a variety of "superwindows" that are very highly insulated. These windows require less heating energy in winter than a well-insulated wall. This is possible because the free heat from sunlight entering the window is greater than the heat lost through the window. This is true even when the window is located on the north wall of a house in Montana.

A different set of technologies provides energy savings in hot climates where prolonged cold winters are of little or no concern. These "cool windows" admit substantial daylight while rejecting most of the sun's heat in the infrared portion of the spectrum.

Researchers are also developing "smart windows" that will be designed to respond to our needs: programmed to automatically modulate the flow of light or solar heat across the window. Experimenters are currently working on various microscopically thin coatings that change their properties in response to heat, light, or electronic signals. The concept of a "smart window" is sometimes explained using the metaphor of a biological cell wall, which actively filters energy and material flows between the inside and outside environments, and changes its filtration rate according to the needs of the cell's metabolism. Smart windows could adjust energy flows to meet the thermal needs of the house, as well as the need for light, view, or privacy.

## WINDOW SELECTION

While individuals make window purchasing decisions based on many unique priorities and circumstances, there are some common considerations that most buyers need to address to make an effective choice. In most cases window buyers base at least part of their decision on appearance factors such as frame style and materials. Then, basic functional issues must be considered. These include providing daylight, glare control, thermal comfort, and ventilation. Certain practical considerations, such as resistance to condensation, sound control, maintenance requirements, and durability, are also part of the decision-making process.

Window energy performance is an important concern, but it is difficult for many designers and homeowners to assess the true impact of choosing a more energy-efficient window. Some of the basic thermal and optical properties (U-factor, solar heat gain coefficient, and air leakage rate) can be identified if the window is properly labeled (see the next section, on NFRC labels). However, consumers still do not know how these basic properties influence annual heating and cooling energy use. This can now be determined by an annual energy rating system or by using various computer tools. Finally, window selection involves cost. This includes not only initial costs but a number of costs and benefits that occur through the life of the window.

One main purpose of this book is to give designers, builders, and homeowners the necessary information to evaluate windows and make intelligent choices. The selection considerations shown at right are discussed in more detail in Chapter 6.

## Window Selection Considerations

**Appearance**
- Size and shape
- Style
- Frame materials
- Glass color and clarity

**Function**
- Daylight
- Glare control
- Fading
- Thermal comfort
- Resistance to condensation
- Ventilation/operating type
- Sound control
- Maintenance
- Durability (warranty)

**Energy Performance**
- Basic energy-related properties
- Annual heating and cooling season performance
- Peak load impacts
- Long-term ability to maintain energy performance

**Cost**
- Initial cost of window units and installation
- Cost of interior and exterior window treatments
- Cost of maintenance
- Frequency of replacement
- Resale value
- Initial cost of heating and cooling system
- Annual cost of heating and cooling energy

## WINDOW RATING SYSTEMS

Until recently, consumers received energy performance information in a variety of ways. Some manufacturers described performance by showing R-values of the glass. While the glass might be a good performer, the rating did not include the effects of the frame. Other manufacturers touted the insulating value of different window components, but these, too, did not reflect the total window system performance.

When manufacturers rated the entire window or other fenestration product, some used test laboratory measurements and others used computer calculations. Even among those using test laboratory reports, the laboratories often tested the windows under different procedures, making an "apples-to-apples" comparison difficult. The different rating methods confused builders and consumers. They also created headaches for manufacturers trying to differentiate their products' performance from their competitors.'

The National Fenestration Rating Council (NFRC) has developed a fenestration energy rating system based on *whole product performance*. This accurately accounts for the energy-related effects of all the products' component parts, and prevents information about a single component from being compared in a misleading way to other whole product prop-

## Window properties rated by the National Fenestration Rating Council (NFRC)

**Basic thermal and optical properties**
- U-value
- Solar heat gain coefficient
- Visible transmittance
- Air infiltration rate

These properties are determined by laboratory tests and detailed simulation procedures. They now appear or will soon appear on NFRC labels.

**Energy performance**
- Annual heating energy
- Annual cooling energy
- Peak heating load
- Peak cooling load

Annual energy performance and peak loads are determined based on simulations using the basic properties. NFRC ratings are a comparative index that is reasonably accurate under most conditions. Annual energy performance ratings will soon appear on labels. Future ratings may include peak heating and cooling loads.

**Other useful properties**
- Condensation resistance
- Long-term energy performance

Work is underway which should lead to NFRC ratings for these properties.

| National Fenestration Rating Council INCORPORATED | |
|---|---|
| **AAA Window Company** | |

| Energy Rating Factors | Ratings | | Product Description |
|---|---|---|---|
| | Residential | Nonresidential | |
| U-factor Determined in accordance with NFRC 100 | 0.40 | 0.38 | Model 1000 Casement Low-e Argon Filled |
| Solar Heat Gain Coefficient Determined in accordance with NFRC 200 | 0.65 | 0.66 | |
| Visible Transmittance Determined in accordance with NFRC 200 | 0.71 | 0.71 | |
| | | | |
| | | | |
| | | | |

*NFRC ratings are determined for a fixed set of environmental conditions and specific product sizes and may not be appropriate for directly determining seasonal energy performance. For additional information contact:*

Figure 1-11. The NFRC label indicates the U-factor, solar heat gain coefficient (SHGC), and the visble transmittance (VT). Future labels will indicate air leakage and annual energy performance ratings.

erties. With energy ratings based on whole product performance, NFRC helps builders, designers, and consumers directly compare products with different construction details and attributes. At this time, NFRC labels on window units give ratings for U-value, solar heat gain coefficient, and visible light transmittance. Soon labels will include air infiltration as well as a Heating Rating (HR) and a Cooling Rating (CR) to indicate annual energy performance. Work is underway which should lead to ratings for other properties, such as condensation resistance and long-term energy performance. The new NFRC logo, shown at right, will be incorporated into a new label design in the near future.

National Fenestration Rating Council

# CHAPTER 2

# Energy Performance Characteristics of Windows

This chapter serves as an introduction to the energy performance of window units. First, there is a brief introduction to the basic mechanisms of heat transfer and how they apply to windows. The next three sections address the key energy-related characteristics of windows–insulating value, ability to control heat gain from solar radiation, and ability to control air leakage. Visible transmittance–a fourth property related to overall window performance—is also described within this chapter.

## HEAT TRANSFER MECHANISMS

Heat flows through a window assembly in three ways: conduction, convection, and radiation. Conduction is heat traveling through a solid material, the way a frying pan warms up. Convection is the transfer of heat by the movement of gases or liquids, like warm air rising from a candle flame. Radiation is the movement of heat energy through space without relying on conduction through the air or by movement of the air, the way you feel the heat of a fire.

When these basic mechanisms of heat transfer are applied to the performance of windows, they interact in complex ways. Thus, conduction, convection, and radiation are not typically discussed and measured separately. Instead, three energy performance characteristics of windows are used to portray how energy is transferred and are the basis for how energy performance is quantified. They are:

- **Insulating value.** When there is a temperature difference between inside and outside, heat is lost or gained through the window frame and glazing by the combined effects of conduction, convection, and radiation. This is indicated in terms of the U-factor of a window assembly.

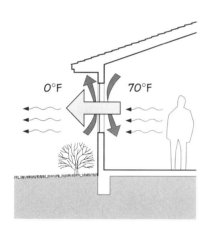

Figure 2-1. This drawing illustrates heat loss through the window in winter by conduction combined with radiation and air movement (convection) on the surfaces of the glazing. The U-factor of a window is a combination of these conductive, convective, and radiative heat transfer mechanisms.

- **Heat gain from solar radiation.** Regardless of outside temperature, heat can be gained through windows by direct or indirect solar radiation. The ability to control this heat gain through windows is measured in terms of the solar heat gain coefficient or shading coefficient of the window glazing.

- **Infiltration.** Heat loss and gain also occur by infiltration through cracks in the window assembly. This effect is measured in terms of the amount of air (cubic feet or meters per minute) that passes through a unit area of window (square foot or meter) under given pressure conditions. In reality, infiltration varies with wind-driven and temperature-driven pressure changes. Infiltration also contributes to summer cooling loads in some climates by raising the interior humidity level.

Each of these energy performance characteristics is described below, along with a brief discussion of approaches to improve energy efficiency in each area. Methods of determining the insulating value, solar heat gain values, and air leakage values of windows are also presented. As previously mentioned, visible transmittance (VT) is an optical property that indicates the amount of visible light transmitted through the glass. Although VT does not directly affect heating and cooling energy use, it is used in the evaluation of energy-efficient windows and therefore is discussed following the solar heat gain section.

## INSULATING VALUE

Heat flow from the warmer side to the colder side of a window and frame is a complex interaction of all three heat transfer mechanisms described above—conduction, convection, and radiation. Figure 2-4 shows the manner in which these basic heat transfer mechanisms interact. The ability of the window assembly to resist this heat transfer is referred to as its *insulating value.* Heat flows from warmer to cooler bodies, thus from inside to outside in winter, and reverses direction in summer during periods when the outside temperature is greater than indoors.

Compared to a well-insulated wall, heat transfer through a typical older window is generally much higher. A single-

Figure 2-2. Solar heat gain passes through glazing to some extent, depending on the glazing type. This gain can be beneficial in winter but undesirable in summer.

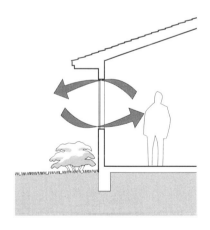

Figure 2-3. Infiltration through cracks in the window assembly is another mechanism for energy transfer.

glazed window has roughly the same insulating qualities as a sheet of metal—most of the insulating value comes from the air layer on each surface of the glass. Such a window can be considered a thermal hole in a wall and typically has a heat loss rate ten to twenty times that of the wall. A window with such a poor insulating value allows heat to flow out of a space almost unimpeded. If the temperature inside is 70°F and outside 0°F (20°C and -18°C), the glass surface of a single-glazed window would be about 17°F (-8°C)—cold enough to form frost on the inside of the glass.

Convection affects the heat transfer in three places in the assembly: the inside glazing surface, the outside glazing surface, and inside any air spaces between glazings. A cold interior glazing surface chills the air adjacent to it. This denser cold air then falls to the floor, starting a convection current. People typically perceive this cold air flow as a "draft" caused by leaky windows, and are tempted to plug any holes they can find, rather than to remedy the situation correctly with a better window that provides a warmer glass surface.

On the exterior, a component of the insulating value of a window is the air film against the glazing surface. As wind blows (convection), this air film is removed or replaced with colder air, which contributes to a higher rate of heat loss. Finally, when there is an air space between layers of glazing, convection currents facilitate heat transfer through this air layer. By adjusting the space between the panes of glass, as well as choosing a gas fill that insulates better than air, double-glazed windows can be designed to minimize this effect.

All objects emit invisible thermal radiation, with warmer objects emitting more than colder ones. Hold your hand in front of an oven window and you will feel the radiant energy emitted by that warm surface. Your hand also radiates heat to the oven window, but since the window is warmer than your hand, the net balance of radiant flow is toward your hand and it is warmed. Now imagine holding your hand close to a single-glazed window in winter. The window surface is much colder than your hand. Each surface emits radiant energy, but since your hand is warmer, it emits more toward the window than it gains and you feel a cooling effect. Thus, a cold glazing surface in a room chills everything else around it.

Through radiant exchange, the objects in the room, and especially the people (who are often the warmest objects), radiate their heat to the colder window. People often feel the

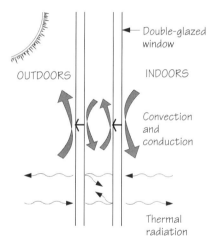

Double-glazed window

INDOORS

OUTDOORS

Convection and conduction

Thermal radiation

Figure 2-4. Components of heat transfer through a window that are related to insulating value.

chill from this radiant heat loss, especially in the exposed skin of their hands and faces, but they attribute the chill to cool room air rather than to a cold window surface. Because of this misconception, few people realize that pulling a curtain across the window as a barrier to the radiant heat loss could be more effective in improving their immediate comfort than turning up the thermostat.

The thermal performance of a roof window or skylight changes according to the angle at which it is mounted. This is because the rate of convective exchange on the inner and outer surfaces as well as within the air space is affected by the slope. Also, skylights and roof windows that point toward the cold night sky have more radiant heat losses at night than windows that view warmer objects such as the ground, adjacent buildings, and vegetation.

## Reducing Winter Heat Loss with Insulating Value

During the nineteenth century, the best protection from the winter cold was storm shutters or heavy drapes, which could help break the force of the winter winds and provide a measure of added insulation, since the single pane windows were intrinsically poor insulators. Around the turn of the century, "storm" windows were conceived as a way to tighten the seal around leaky, rattling window sashes during the winds of a winter storm, as well as to create a layer of insulating air between two glazings. Since they restricted the use of the window for ventilation, they had to be installed every winter and removed every spring. Later innovations provided sliding storm and screen sashes in the same unit, which required no changing of windows.

More recently, a series of new strategies and technological developments have led to substantial improvement in the insulating value of windows. These include increasing the number of glazing layers, using insulating gas fills in the air spaces, using low-emittance coatings to reduce radiant heat loss, and improving the insulating value of glazing edge spacers, sashes, and frames. These improvements, which can significantly increase the insulating value of the entire window assembly, are explained in detail in Chapter 3.

## Reducing Summer Heat Gain with Insulating Value

Insulating glazing can also reduce the flow of heat *into* a house in warm climates. However, the difference between indoor and outdoor temperatures tends to be considerably less in a cooling situation than in a heating one. Because this smaller temperature differential creates less of a driving force for heat transfer, improving the insulating value to reduce heat flow during warm weather tends to be of secondary importance. For this reason, insulating windows have had a slower acceptance rate in the warmer parts of the country, which have short heating seasons and long cooling seasons. In 1994, approximately 95 percent of houses in the northeastern United States had storm sash or insulating glass windows, while only 55 percent of houses in the southeastern United States had such windows.

Even though fewer warm climate houses use insulating windows, the number is increasing due to the growing prevalence of residential air conditioning. The amount of money spent to air-condition a home can often exceed the cost of heating it, largely because the fuel source for air conditioning, electricity, is typically so much more expensive than the fuels generally used for heating. This economic factor has motivated the use of better insulating windows in hot climates. The same strategies listed above to improve the insulating value of windows to reduce winter heat loss are applicable to reducing summer gain as well.

## Determining Insulating Value

The U-factor (also referred to as U-value) is the standard way to quantify insulating value. It indicates the rate of heat flow through the window. The U-factor is the total heat transfer coefficient of the window system (in Btu/hr-sq ft-°F or W/sq m-°C), which includes conductive, convective, and radiative heat transfer. It therefore represents the heat flow per hour (in Btus per hour or watts) through each square foot (or square meter) of window for a 1°F (1°C) temperature difference between the indoor and outdoor air temperature. The R-value is the reciprocal of the total U-factor (R=1/U). As opposed to an R-value, the smaller the U-factor of a material, the lower the rate of heat flow.

In addition to the thermal properties of the materials in the window assembly, the U-factor depends on the weather conditions, such as the temperature differential between indoors and out, and wind speed. Window manufacturers typically list a U-factor for winter that is determined under relatively harsh conditions: 15 mph (25 km/hr) wind, 70°F (20°C) indoors, 0°F (-18°C) outdoors. These conditions have been standardized with a set temperature and wind speed so that product ratings can be used for comparison purposes.

The U-factor of a total window assembly is a combination of the insulating values of the glazing itself, the edge effects that occur in the insulated glazing unit, and the window frame and sash.

## Glass

The U-factor of the glazing portion of the window unit is affected primarily by the total number of glazing layers, the dimension separating the various layers of glazing, the type of gas that fills the separation, and the characteristics of coatings on the various surfaces. The U-factor for the glazing alone is referred to as the center-of-glass U-factor (Figure 2-5).

## Edge Effects

A U-factor calculation assumes that heat flows perpendicular to the plane of the window. However, windows are complex three-dimensional assemblies, in which materials and cross sections change in a relatively short space.

For example, metal spacers at the edge of an insulating glass unit have much higher heat flow than the center of the insulating glass, which causes increased heat loss along the outer edge of the glass. The relative impact of these "edge effects" becomes more important as the insulating value of the rest of the assembly increases.

## Frames and Sashes

The heat loss through a window frame can be quite significant: in a typical four-foot by three-foot (1.2 by 0.9 m) double-hung wood frame window, the frame and sash can occupy approximately 30 percent of the window area.

In a frame with a cross section made of one uniform, solid material, the U-factor is based on the conduction of heat through the material. However, hollow frames and composite frames with various reinforcing or cladding materials are

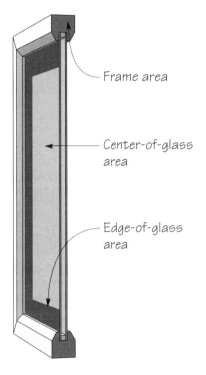

Frame area

Center-of-glass area

Edge-of-glass area

Figure 2-5. Principle zones for determining heat loss through a window assembly.

more complex. Here, conduction through materials must be combined with convection of the air next to the glazing and radiant exchange between the various surfaces.

Furthermore, window frames rarely follow the same cross section around a window. For example, a horizontal slider has seven different frame cross sections, each with its own rate of heat flow. Because of these complications, determining the U-factor through a window frame is very complex and is best calculated by specialized computer programs or determined by test.

## Overall U-factor

Since the U-factors are different for the glass, edge-of-glass zone, and frame, it can be misleading to compare U-factors if they are not carefully described. In order to address this problem, the concept of a total window U-factor is utilized by the National Fenestration Rating Council (NFRC). A specific set of engineering assumptions and procedures must be followed to calculate the overall U-factor of a window unit using the NFRC method. Figure 2-6 indicates the center-of-glass U-factor and the overall U-factor for several types of window unit. In most cases, the overall U-factor is higher than the U-

Figure 2-6. U-factors of various window assemblies.

| Window description | Center-of-glass U-factor (Btu/hr-ft$^2$-°F) | Overall U-factor (Btu/hr-ft$^2$-°F) |
|---|---|---|
| 1  Single glass<br>Aluminum frame<br>with no thermal break | 1.11 | 1.30 |
| 5  Double glass<br>Wood /vinyl frame | 0.49 | 0.49 |
| 9  Double glass<br>Selective low-E (.04)–argon<br>Wood /vinyl frame | 0.24 | 0.29 |
| 11 Triple glass<br>Low-E (2 surfaces)–krypton<br>Insulated vinyl frame | 0.11 | 0.15 |

Note: Windows are described in Appendix A.

factor for the glass alone, since the glass remains superior to the frame in insulating value.

The U-factor of a window unit is typically described for the unit in a vertical position. A change in mounting angle can affect its U-factor. A roof window in a vertical position might have a U-factor of 0.50; the same unit installed on a sloped roof at 20 degrees from horizontal would have a U-factor of 0.56 (12 percent higher under winter conditions).

## Energy Effects of U-factor

Figure 2-7 illustrates the effect of changing the window U-factor on annual energy use in three climates. Even though the total window area is a small percentage of the total building envelope, the impact of changing the U-factor can be a significant factor in the overall building performance.

Figure 2-7. Effect of U-factor on heating season energy performance.

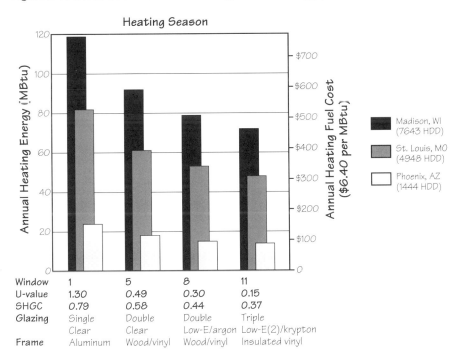

Note: The annual energy performance figures shown here are for a typical 1540 sq ft house. U-factor and SHGC are for total window including frame. House and windows are described in Appendix A. MBtu=millions of Btu. HDD=heating degree days.

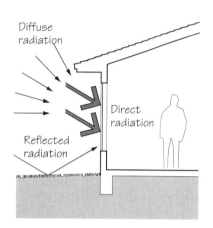

Figure 2-8. Solar heat gain includes direct, reflected, and diffuse radiation.

## SOLAR RADIATION CONTROL

The second major energy performance characteristic of windows is the ability to control solar heat gain through the glazing. Solar heat gain through windows tends to be the single most significant factor in determining the air-conditioning load of a residential building. The intensity of heat gain from solar radiation can greatly surpass heat gain from other sources, such as outdoor air temperature or humidity.

The origin of solar heat gain is the direct and diffuse radiation coming directly from the sun and the sky or reflected from the ground and other surfaces. Some radiation is directly transmitted through the glazing to the space, and some may be absorbed in the glazing and then indirectly admitted to the space. Other thermal (nonsolar) heat transfer effects are included in the U-factor of the window. Sunlight is composed of electromagnetic radiation of many wavelengths, ranging from short-wave invisible ultraviolet, to the visible spectrum, to the longer, invisible near-infrared waves. About half of the sun's energy is visible light; the remainder is largely infrared with a small amount of ultraviolet. This characteristic of sunlight makes it possible to selectively admit or reject different portions of the solar spectrum. While reducing solar radiation through windows is a benefit in some climates and during some seasons, maximizing solar heat gain can be a significant energy benefit under winter conditions. These often conflicting directives can make selection of the "best" window a challenging task.

### Maximizing Solar Radiation

During the heating season, the heat contained in the sun's radiation that enters windows can be just as useful as the light. Houses typically benefit from solar heat gain in the winter, when the sun takes the lowest and shortest arching path through the sky. Rising south of east and setting south of west (in the northern hemisphere), the sun will shine into a south window essentially all day during the winter. Thus, increasing the proportion of windows facing south is a direct, and usually no-cost, way to benefit from the sun's heat. A well-insulated house with properly oriented windows may often require no supplemental heat during sunny winter days.

In order to maximize winter heat gain from solar radiation, the following strategies can be applied:

- Use clear glass (special low-iron glass can be used to further improve solar transmission).

- Balance the insulating benefits of additional glazings against reduced solar gain through the additional layers. For example, each additional layer of clear glazing can reduce the amount of solar radiation by about 14 percent.

- If low-emittance coatings are used to increase the insulating value of the windows, select coatings that maximize incoming solar gain.

- Place materials with a high heat capacity, such as masonry, in the path of the sunlight. The materials soak up the excess heat while the sun is shining and re-émit it later when needed.

Some of these strategies may be in conflict with the need to reduce summer cooling loads, as explained below.

## Minimizing Solar Radiation

Traditionally, overhangs, awnings, trees, and other external shading devices have been the best strategy to minimize solar radiation by blocking it before it reaches the window. Once the sun actually strikes the window surface, reducing the solar heat gain is a more difficult task. Before relatively recent innovations such as tints and coatings, residential windows were limited to one or two layers of 1/8-inch (3 mm) clear glass, which allows 75 to 85 percent of the solar energy to enter. By adding interior curtains or blinds, some of the sunlight is reflected back out through the glass, but most of the sun's energy remains inside.

The following window design strategies are used to minimize solar radiation:

- Use tinted glass. Various shades of tinted glass reduce the solar gain by 25 to 55 percent. Some tints block the full solar spectrum—both infrared heat as well as visible light, while others are "spectrally selective"—they transmit daylight while absorbing and blocking near-infrared radiation. A problem with single-pane tinted glass is that it blocks solar radiation mostly by absorption. This makes the glass heat up, and much of this heat enters the building, thus diminishing the net impact of the glazing on solar heat gain reduction.

- Use reflective coatings. A reflective coating on glass can be altered in density, to reflect just a little or most of the light. Reflective coatings can be applied to clear or tinted glass and can reduce heat gain by 60 to 80 percent—but they also reduce light transmission by 70 to 90 percent. If the reflective coating is on the outside, less heat is absorbed by the glass and transferred to the interior.

- Use spectrally selective reflective coatings. Most reflective glazings that are effective in reducing solar radiation also greatly reduce the amount of transmitted daylight. Spectrally selective coatings are designed to allow the most beneficial portions of sunlight (visible daylight) through the glass and selectively reflect the undesirable portions (ultraviolet and near infrared). These coatings are also low-emittance, so they have additional insulating benefits.

Recent developments such as low-emittance and spectrally selective coatings are explained in detail in Chapter 3.

## Determining Solar Heat Gain

There are two means of indicating the amount of solar radiation that passes through a window. These are solar heat gain coefficient (SHGC) and shading coefficient (SC). In both cases, the solar heat gain is the combination of directly transmitted radiation and the inward-flowing portion of absorbed radiation (Figure 2-9). However, SHGC and SC have a different basis for comparison or reference.

### Solar Heat Gain Coefficient

The window industry is now moving away from use of the shading coefficient to the solar heat gain coefficient (SHGC), which is defined as that fraction of incident solar radiation that actually enters a building through the window as heat gain. The SHGC generally refers to total window system performance and is a more accurate indication of solar gain under a wider range of conditions. The solar heat gain coefficient is expressed as a dimensionless number from 0 to 1.0. A high coefficient signifies high heat gain, while a low coefficient means low heat gain. Typical SHGC values for both the total window and the glass alone are shown in Figure 2-10.

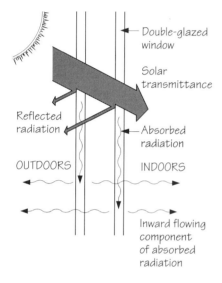

Double-glazed window

Solar transmittance

Reflected radiation

Absorbed radiation

OUTDOORS    INDOORS

Inward flowing component of absorbed radiation

Figure 2-9. Simplified view of the components of solar heat gain. Heat gain includes the transmitted solar energy and the inward flowing component of absorbed radiation.

## Shading Coefficient

Until recently, the shading coefficient (SC) was the primary term used to characterize the solar control properties of glass in windows. Although it is being replaced by the solar heat gain coefficient, you will still find it referenced in books or product literature. The SC was originally developed as a single number that could be used to compare glazing solar control under a wide range of conditions. Its simplicity, however, is offset by a lack of accuracy in a number of circumstances.

The shading coefficient (SC) is only defined for the glazing portion of the window and does not include frame effects. It represents the ratio of solar heat gain through the system

Figure 2-10. Solar heat gain and visible transmittance characteristics of typical windows.

| | Window | | | | | |
|---|---|---|---|---|---|---|
| | 1 | 5 | 6 | 9 | 10 | 11 |
| General glazing description | Single-glazed clear | Double-glazed clear | Double-glazed bronze | Double-glazed spectrally selective | Double-glazed spectrally selective | Triple-glazed low-E (2) |
| **Center-of-glass** | | | | | | |
| SHGC | 0.86 | 0.76 | 0.62 | 0.41 | 0.32 | 0.49 |
| SC | 1.00 | 0.89 | 0.72 | 0.47 | 0.38 | 0.57 |
| VT | 0.90 | 0.81 | 0.61 | 0.72 | 0.44 | 0.68 |
| LSG | 1.04 | 1.07 | 0.98 | 1.75 | 1.38 | 1.39 |
| **Total window** | | | | | | |
| SHGC | 0.79 | 0.58 | 0.48 | 0.31 | 0.26 | 0.37 |
| VT | 0.69 | 0.57 | 0.43 | 0.51 | 0.31 | 0.48 |
| LSG | 0.87 | 0.98 | 0.89 | 1.65 | 1.19 | 1.29 |

Note:  Windows are described in Appendix A. Values for total window are based on a 2 foot by 4 foot casement window.

SHGC = solar heat gain coefficient
SC = shading coefficient
VT = visible transmittance
LSG = light-to-solar-gain ratio
    (VT/SHGC)

relative to that through 1/8-inch (3 mm) clear glass at normal incidence. The SC has also been used to characterize performance over a wide range of sun positions; however, there is some potential loss in accuracy when applied to sun positions at high angles to the glass. The shading coefficient is expressed as a dimensionless number from 0 to 1.0. A high shading coefficient means high solar gain, while a low shading coefficient means low solar gain.

The total window SC value is strongly influenced by the type of glass selected. The shading coefficient can also include the effects of any integral part of the window system that reduces the flow of solar heat, such as multiple glazing layers, reflective coatings, or blinds between layers of glass. The SHGC is influenced by all the same factors as the SC, but since it can be applied to the entire window assembly, the SHGC is also affected by shading from the frame as well as the ratio of glazing and frame. Note in Figure 2-10 that for any glazing, the SHGC is always lower than the SC. If you find an older information source with SC values only, you can make an approximate conversion to SHGC for the glazing by multiplying the SC value by 0.87. Total window SHGC is used throughout this book and is found on the NFRC window labels.

### Energy Effects of SHGC

Figure 2-11 illustrates the effect of changing the solar heat gain coefficient on cooling energy use in three climates. In a predominantly cooling climate such as Phoenix, Arizona, reducing the SHGC results in a noticeable decrease in the cooling season energy use.

## VISIBLE TRANSMITTANCE

Visible transmittance is the amount of light in the visible portion of the spectrum that passes through a glazing material. This property does not directly affect heating and cooling loads in a building, but it is an important factor in evaluating energy-efficient windows. Transmittance is influenced by the glazing type, the number of layers, and any coatings that might be applied to the glazings. These effects are discussed in more detail in Chapter 3 in conjunction with a review of various glazing and coating technologies. Visible transmittance of glazings ranges from above 90 percent for water-white

Figure 2-11. Effect of solar heat gain coefficient (SHGC) on annual cooling season energy performance.

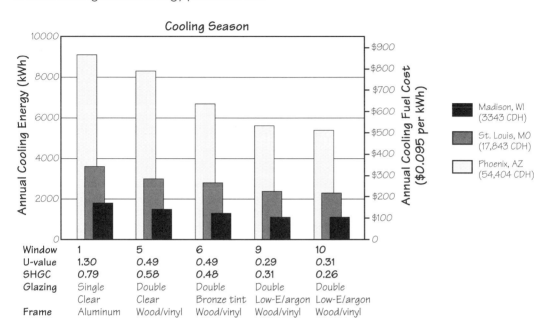

| Window | 1 | 5 | 6 | 9 | 10 |
|---|---|---|---|---|---|
| U-value | 1.30 | 0.49 | 0.49 | 0.29 | 0.31 |
| SHGC | 0.79 | 0.58 | 0.48 | 0.31 | 0.26 |
| Glazing | Single Clear | Double Clear | Double Bronze tint | Double Low-E/argon | Double Low-E/argon |
| Frame | Aluminum | Wood/vinyl | Wood/vinyl | Wood/vinyl | Wood/vinyl |

Note: The annual energy performance figures shown here are for a typical 1540 sq ft house. U-factor and SHGC are for total window including frame. House and windows are described in Appendix A. kWh=kilowatt hours. CDH=cooling degree hours.

clear glass to less than 10 percent for highly reflective coatings on tinted glass.

Visible transmittance is an important factor in providing daylight, views, and privacy, as well as in controlling glare and fading of interior furnishings. These are often contradictory effects: a high light transmittance is desired for view out at night, but this may create glare at times. These opposing needs are often met by providing glazing that has high visible transmittance and then adding window attachments such as shades or blinds to modulate the transmittance to meet changing needs.

NFRC now reports visible transmittance as a rating on its label. Note that NFRC's rating is a whole window rating that combines the effect of both glazing and window frame. There are many cases where the transmittance of glazing alone will be required, so it is important to make sure that the appropriate properties are being compared.

In the past, products that reduced solar gain (with tints and coatings) also reduced visible transmittance. However, new spectrally selective tinted glasses and selective coatings have made it possible to reduce solar heat gain with little reduction in visible transmittance. Because the concept of separating solar gain control and light control is so important, we now characterize this differential effect by referring to the light-to-solar-gain ratio (LSG), which is simply the ratio of visible transmittance to solar heat gain coefficient. Values for typical glazings are listed in Figure 2-10.

## INFILTRATION

Infiltration can be defined as ventilation that is not controlled and usually not wanted. It is the leakage of air through cracks in the building envelope. Infiltration leads to increased heating or cooling loads when the outdoor air entering the building needs to be heated or cooled. Windows and doors are typically responsible for a significant amount of the infiltration in homes. In extreme conditions, depending on the window type and quality, infiltration can be responsible for as much heat loss or gain as the rest of the window. Tight sealing and weatherstripping of windows, sash, and frames is of paramount importance in controlling infiltration.

The use of good-quality fixed windows helps to reduce infiltration because these windows are easier to seal and keep tight. Operable windows are necessary for ventilation, but they are also more susceptible to air leakage. Operable window units with low air leakage rates are characterized by good design and high-quality construction and weatherstripping. They also feature mechanical closures that positively clamp the window shut against the wind. For this reason, compression-seal windows such as awning, hopper, and casement designs are generally more effectively weatherstripped than are sliding-seal windows. Sliding windows rely on wiper-type weatherstripping, which is more subject to wear over time and can be bypassed when the window flexes under wind pressure.

Figure 2-13 illustrates the effect of changes in window air leakage on annual energy use in a cold climate. Most industry groups recommend air leakage values of 0.56 cfm/sq ft or lower. Even though window air leakage is not as significant as insulating value or control of solar radiation, it still can have

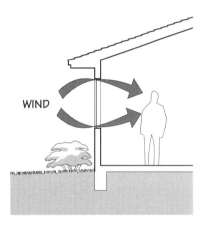

WIND

Figure 2-12. The pressure difference created by wind is a major influence on infiltration through cracks in the window assembly. Wind exposure is determined by house location and site design.

a noticeable negative effect if windows are not tight.

The level of infiltration depends upon local climate conditions, particularly wind conditions and microclimates surrounding the house. Wind effects increase rapidly as you move away from the protection of trees, shrubs, or other buildings. Infiltration can be a significant issue in heating costs, especially where winter temperature differentials between inside and outside are quite high and during windy weather conditions. Infiltration generally plays a much lesser role relative to cooling cost, because the difference between indoor and outdoor temperatures tends to be lower and accompanied by milder winds. In very humid locations, however, infiltration can introduce a large latent cooling load.

Figure 2-13. Effect of window air leakage rate on annual energy performance. Most manufacturers produce windows rated at 0.56 cfm/sq ft or below. Higher leakage rates represent older windows or windows without weatherstripping.

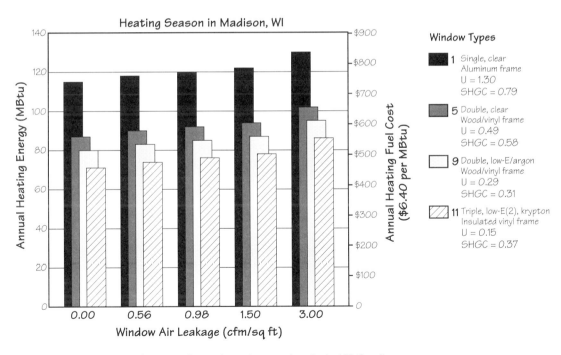

Note: The annual energy performance figures shown here are for a typical 1540 sq ft house (see Appendix A). The windows modeled in this figure are 2-foot by 4-foot casements. MBtu=millions of Btu, cfm/sq ft=cubic feet per minute per square foot of window area.

Cracks and air spaces left between the window unit and the building wall can also account for considerable infiltration. Insulating and sealing these areas during construction or renovation can be very effective in controlling infiltration.

## Measuring Air Leakage

The amount of air that leaks through all of the cracks in and around a window sash and frame is a function of crack length, tightness of the seals and joints, and the pressure differential between the inside and outside. American window manufacturers report air leakage test values either as cfm/sq ft (cubic feet per minute per square foot of window area) or cfm/lf (cubic feet per minute per lineal foot of sash crack). The NFRC rating system requires that air leakage be reported in cfm per square foot for all products so that performance can be compared across product types.

According to a standard procedure referred to as ASTM E-283 (see ASTM in Appendix D), windows are tested for the amount of air that will pass through a window unit, excluding the air that passes between the frame of the unit and the wall. The unit is closed and locked during testing. The unit is tested under a static air pressure of 1.56 pounds per square foot, equivalent to a 25 mph wind velocity. Some high-performance windows are tested at a static pressure of 6.27 pounds per square foot, equivalent to a 50 mph wind velocity.

The manufacturer's test may be for a typical unit or for an average of unit sizes. It is important to note that the measured leakage number reflects performance under laboratory conditions, not installed conditions. Rough handling or damage during installation or use tends to lower the performance of the window. Windows also tend to have decreased performance, i.e., greater air leakage rates, with general age and use as seals and weatherstripping deteriorate.

# CHAPTER 3

# Window Glazing Materials

The remarkable advances in the performance of windows in recent years are based mainly on technological developments in glazing materials. The first section of the chapter defines key properties of glazing that affect energy performance: transmittance, reflectance, absorptance, and emittance. The next sections describe standard and advanced glazing materials that are currently on the market. Then the ongoing developments of specialty and experimental glazings are discussed. Finally, we look at the use of thermograms to illustrate relative window performance.

The overall energy performance of a window assembly is dependent not only on the glazing and frame materials themselves, but on a number of architectural and interior design decisions. These include exterior features such as overhangs,

Figure 3-1. Vacuum chambers are used in the production of sputtered low-emittance glass.

awnings, screens, and other shading devices. On the interior, drapes, blinds, shades, and movable insulation all contribute to the overall performance. These exterior and interior additions to the windows are described in Chapter 6. It is interesting to note that as more advanced glazing products are used, the energy-related impact and importance of these traditional control strategies are diminished, since the glazing itself is providing significant solar control and thermal resistance.

## PROPERTIES OF GLAZING MATERIALS RELATED TO RADIANT ENERGY TRANSFER

Three things happen to solar radiation as it passes through a glazing material. Some is transmitted, some is reflected, and the rest is absorbed. These are the three components of solar impact that determine many of the other energy performance properties of a glazing material, such as the solar heat gain coefficient and shading coefficient discussed in Chapter 2. Manipulating the proportion of transmittance, reflectance, and absorptance for different wavelengths of solar radiation has been the source of much recent innovation in window energy performance.

Before the technical innovations available today existed, the primary characteristic of glass was its ability to transmit visible light—in other words, how clear it was. However, as attention focused on improving the total energy performance of glass, it became clear that transparency to visible light is only part of the picture.

Visible light is a small portion of the electromagnetic spectrum (Figure 3-3). Beyond the blues and purples lie ultraviolet radiation and other higher-energy short wavelengths, from X rays to gamma rays. Beyond red light are the near-infrared, given off by very hot objects, the far-infrared, given off by warm room-temperature objects, and the longer microwaves and radio waves.

Glazing types vary in their transparency to different parts of the spectrum. On the simplest level, a glass that appears to be tinted green as you look through it toward the outside will transmit more sunlight from the green portion of the visible spectrum, and reflect/absorb more of the other colors. Similarly, a bronze-tinted glass will absorb the blues and greens and transmit the warmer colors. Neutral gray tints absorb most colors equally.

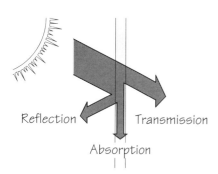

Figure 3-2. The passage of solar radiation through a glazing material.

This same principle applies outside the visible spectrum. Most glass is partially transparent to at least some ultraviolet radiation, while plastics are commonly more opaque to ultraviolet. Glass is opaque to far-infrared radiation but generally transparent to near-infrared. Strategic utilization of these variations has made for some very useful glazing products.

The four basic properties of glazing that affect radiant energy transfer are transmittance, reflectance, absorptance, and emittance. Each is described below, and their application in developing new products is presented later in the chapter.

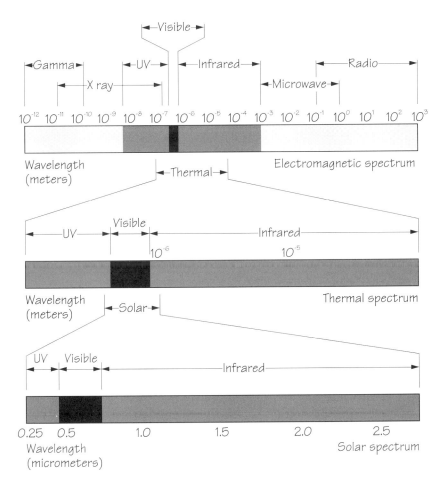

Figure 3-3. The electromagnetic spectrum.

Note the scale changes between the three charts.

## Transmittance

Transmittance refers to the percentage of radiation that can pass through glazing. Transmittance can be defined for different types of light or energy, e.g., "visible light transmittance," "UV transmittance," or "total solar energy transmittance." Each describes a different characteristic of the glazing.

Transmission of visible light determines the effectiveness of a type of glass in providing daylight and a clear view through the window. For example, tinted glass has a lower visible light transmittance than clear glass. While the human eye is sensitive to light at wavelengths of about 0.4 to 0.7 micrometers, its peak sensitivity is at 0.55, with lower sensitivity at the red and blue ends of the spectrum. This is referred to as the photopic sensitivity of the eye.

More than half of the sun's energy is invisible to the eye and reaches us as either ultraviolet (UV) or, predominantly, as near-infrared (Figure 3-4). Thus, "total solar energy transmittance" describes how the glazing responds to a much broader part of the spectrum and is more useful in characterizing the quantity of solar energy transmitted by the glazing.

Ultraviolet radiation fades furniture, drapes, and carpets. Therefore, a reduction in the ultraviolet transmittance is typically considered a benefit. We see this in advertisements for sunglasses, but the same principles apply to windows. As noted above, most glass types are at least partially transparent to ultraviolet, while plastics tend to be more opaque to UV. Since visible light plays some role in fading as well, eliminating UV transmittance alone does not provide complete fade protection.

With the recent advances in glazing technology, manufacturers can control how glazing materials behave in these different areas of the spectrum. The basic properties of the substrate material (glass or plastic) can be altered, and coatings can be added to the surfaces of the substrates. For example, a window optimized for daylighting and for reducing heat gains should transmit an adequate amount of light in the visible portion of the spectrum, while excluding unnecessary heat gain from the near-infrared part of the electromagnetic spectrum.

On the other hand, a window optimized for collecting solar heat gain in winter should transmit the maximum amount of visible light as well as the heat from the near-infrared wavelengths in the solar spectrum, while blocking the lower-energy

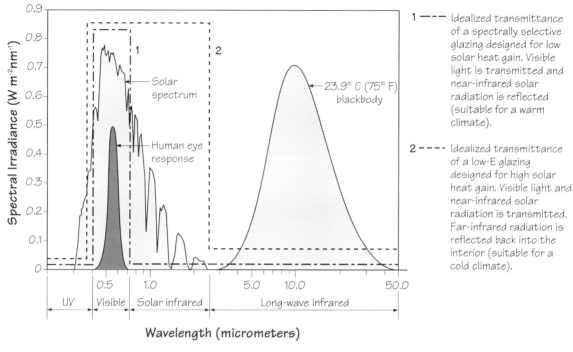

Figure 3-4. Ideal spectral transmittance for glazings in different climates. (Source: McCluney, 1996.)

radiant heat in the far-infrared range that is an important heat loss component. These are the strategies of spectrally selective and low-emittance coatings, described later in the chapter.

## Reflectance

Just as some light reflects off of the surface of water, some light will always be reflected at every glass surface. A specular reflection from a smooth glass surface is a mirror-like reflection similar to when you see an image of yourself in a store window. The natural reflectivity of glass is dependent on the quality of the glass surface, the presence of coatings, and the angle of incidence of the light. Today, virtually all glass manufactured in the United States is float glass and has a very similar quality with respect to reflectance. The sharper the angle at which the light strikes, however, the more the light is reflected rather than transmitted or absorbed (Figure 3-5). Even clear glass reflects 50 percent or more of the light striking it at incident angles greater than about 70 degrees.

(The incident angle is formed with respect to a line perpendicular to the glass surface.)

Coatings can often be detected by careful examination of a reflected bright image, even if the coating is a transparent low-E coating. Hold a match several inches from a window at night and observe the reflections of the match in the glass. You will see two closely spaced images for each layer of glass, since the match reflects off the front and back surface of each layer of glass. A wider spacing between the two sets of pairs of images occurs with a wider air space between the glass panes. A subtle color shift in one of the reflected images normally indicates the presence of a low-E coating.

The reflectivity of various glass types becomes especially apparent during low light conditions. The surface on the brighter side acts like a mirror because the amount of light

Figure 3-5. Light transmitted and reflected by 1/8-inch (3 mm) clear glass as a function of the incident angle.

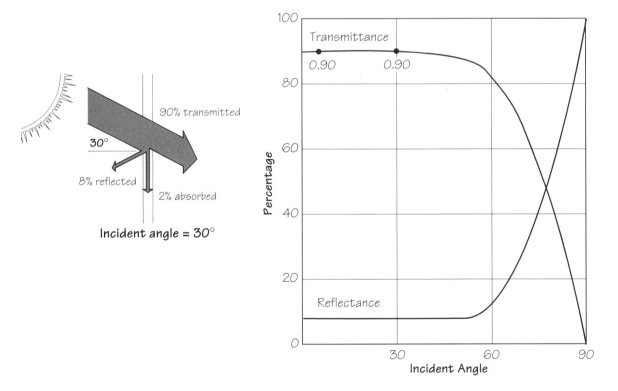

passing through the window from the darker side is less than the amount of light being reflected from the lighter side. This effect can be noticed from the outside during the day and from the inside during the night. For special applications when these surface reflections are undesirable (i.e., viewing merchandise through a store window on a bright day), special coatings can virtually eliminate this reflective effect.

The reflectivity of glass can be increased by applying various metallic coatings to the surface. Early processes used a liquid alloy of mercury and tin to create mirrors. A silvering process developed in 1865 improved the performance of mirrors. Today, mirror-like surfaces can be created by using vacuum-deposited aluminum or silver, or with a durable pyrolytic coating applied directly to the glass as it is manufactured. Thick coatings can be fully reflective and virtually opaque; a thinner coating is partially reflective and partially transmitting. "One-way glass" has such a partial metallic coating, which acts like a full mirror when viewed from the side with brighter light conditions and yet appears transparent when viewed from the side with dimmer light. This simple technology for the production of partial mirrors also led to the creation of more sophisticated reflective coatings for window glass.

Most common coatings reflect all portions of the spectrum. However, in the past twenty years, researchers have learned a great deal about the design of coatings that can be applied to glass and plastic to reflect only selected wavelengths of radiant energy. Varying the reflectance of far-infrared and near-infrared energy has formed the basis for low-emittance coatings for cold climates, and for spectrally selective low-emittance coatings for hot climates.

## Absorptance

Energy that is not transmitted through the glass or reflected off of its surfaces is absorbed. Once glass has absorbed any radiant energy, the energy is transformed into heat, raising the temperature of the glass.

Typical 1/8-inch (3 mm) clear glass absorbs only about 4 percent of incident sunlight. The absorptance of glass is increased by adding to the glass chemicals that absorb solar energy. If they absorb visible light, the glass appears dark. If they absorb ultraviolet radiation or near-infrared, there will

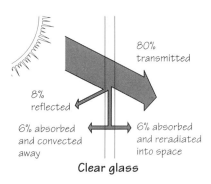

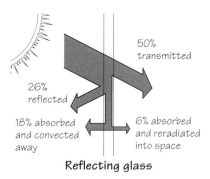

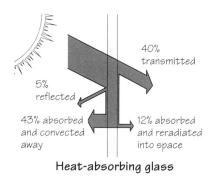

Figure 3-6. Transmission through three types of glass.

be little or no change in visual appearance. Clear glass absorbs very little visible light, while dark tinted glass absorbs a considerable amount. The absorbed energy is converted into heat, warming the glass. Thus, when these "heat-absorbing" glasses are in the sun, they feel much hotter to the touch than clear glass. They are generally gray, bronze, or blue-green and are used primarily to lower the solar heat gain coefficient and to control glare. Since they block some of the sun's energy, they reduce the cooling load placed on the building and its air-conditioning equipment. Absorption is not the most efficient way to reduce cooling loads, as discussed later.

All glass and most plastics, however, are generally very absorptive of far-infrared energy. This property led to the use of clear glass for greenhouses, where it allowed the transmission of intense solar energy but blocked the retransmission of the low-temperature heat energy generated inside the greenhouse and radiated back to the glass.

## Emittance

When heat or light energy is absorbed by glass, it is either convected away by moving air or reradiated by the glass surface. This ability of a material to radiate energy is called its *emissivity*. Windows, along with all other household objects, typically emit, or radiate, heat in the form of long-wave far-infrared energy. This emission of radiant heat is one of the important heat transfer pathways for a window. Thus, reducing the window's emission of heat can greatly improve its insulating properties.

Standard glass has an emittance of 0.84 over the long wavelength portion of the spectrum, meaning that it emits 84 percent of the energy possible for an object at its temperature. It also means that for long-wave radiation striking the surface of the glass, 84 percent is absorbed and only 16 percent is reflected. By comparison, low-E glass coatings have an emittance as low as 0.04. This glazing would emit only 4 percent of the energy possible at its temperature, and thus reflect 96 percent of the incident long-wave infrared radiation.

## BASIC GLAZING MATERIALS

Two basic materials are used for window glazing: glass, which is by far the most common, and plastics, which have many specialized applications.

### Glass

Traditionally, windows have been made of clear glass. The earliest clear glass, produced largely from available sand and formed into sheets with glass-blowing techniques, tended to have a greenish color, entrained bubbles, and an uneven surface. As technology has advanced, the glass color has become "whiter," there are no longer visible impurities, and the surfaces have become more parallel and polished.

Most residential-grade clear glass today is produced with the float technique, described in Chapter 1, in which the glass is "floated" over a bed of molten tin. This provides extremely flat surfaces, uniform thicknesses, and few if any visual distortions. The glass has a slight greenish cast, due to iron impurities, but this is generally not noticeable except from the edge. An even higher-quality glass with reduced iron content eliminates the greenness and also provides a higher solar energy transmittance. This is commonly called "water-white glass."

Obscure glasses still transmit most of the light but break up the view in order to provide privacy, as in bathroom windows. This effect is generally achieved either with decorative embossed patterns or with a frosted surface that scatters the light rays.

By adding various chemicals to glass as it is made, glass can be produced in a wide variety of colors, as we know from stained glass. Glass colors are typically given trade names, but the most frequently used colors can be generally described as clear, bronze, gray, and blue-green. After clear glass, the gray glasses are most commonly used in residential construction, as they have the least effect on the perceived color of the light. Tinted glass is discussed later in the chapter.

The mechanical properties of glass can be altered, as well as its basic composition and surface properties. Heat-strengthening and tempering make glass more resistant to breakage. Heat-strengthened glass is about twice as strong as standard glass. Tempered glass is produced by reheating and then

Figure 3-7. Tempered glass is required in sliding glass doors.

quickly chilling the glass. It breaks into small fragments, rather than into long, possibly dangerous shards. Laminated glass is a sandwich of two outer layers of glass with a plastic inner layer that holds the glass pieces together in the event of breakage. Fully tempered and laminated glass is required by building codes in many door and window applications.

Advanced technologies that have significantly improved the thermal performance of glass are discussed below, in the section on Improved Glazing Products.

## Plastics

Several plastic materials have been adapted for use as glazing materials. Their primary uses are windows with special requirements and skylights.

The following list of plastic glazing materials does not necessarily cover every commercially available product, but it indicates the major types of plastic glazing materials and compares their general properties:

- Clear acrylic is widely available and relatively inexpensive. It is available in various tints and colors. It has excellent visible light transmittance and longevity. However, it is softer than glass, which makes it vulnerable to scratching.

- Frosted acrylic is like clear acrylic, except that it diffuses light and obscures the view. It comes in varying degrees of light transmittance. Most bubble skylights are made of frosted acrylic.

- Clear polycarbonate is like acrylic sheet, but it is harder and tougher, offering greater resistance to scratching and breakage. It is more expensive than acrylic.

- Fiber-reinforced plastic is a tough, translucent, flexible sheet material with good light-diffusing properties. Short lengths of fiberglass are embedded in a polymer matrix to form flat or ribbed sheets. Stiff, insulating, translucent panels are created by bonding double layers to a metal frame and adding fiberglass insulation. It is also formed into corrugated sheets as a translucent roofing material. Surface erosion may shorten its useful life.

- Extruded multicell sheet, usually made with acrylic or polycarbonate plastic, is a transparent or tinted plastic extruded into a double- or triple-wall sheet with divider webs for stiffness, insulating value, and light diffusion.

- Polyester is a thin film used to carry specialized coatings and/or to divide the air space between two layers of glass into multiple air spaces. Highly transparent, it is protected from abuse and weathering by the two exterior glass layers. It can also be used in tinted or coated forms as film that is glued to the inner surface of existing windows for retrofitting applications.

---

**Comparison of Glass and Plastic Materials**

- Glass is extremely durable and essentially maintenance free. The surface and/or transparency of plastic may degrade with time and exposure to moisture and sunlight.

- Glass has lower thermal expansion than plastic and can readily tolerate higher or lower operating temperatures.

- Glass is impermeable to gases and moisture and can thus be used with hermetically sealed, gas-filled insulating units. Plastics are not normally fabricated in sealed units.

- Glass and plastic can be fabricated with tints and light-diffusing properties. Rigid plastics are not normally coated, although plastic films are widely coated with low-temperature processes; glass can be coated easily by a variety of high- and low-temperature processes.

- Plastics are less brittle than glass and do not shatter when broken. Tempered glass reduces injury potential when glass breaks. Plastics are tougher and can resist vandals. Conventional glass can be laminated with plastics to produce a wide range of burglar- and bullet-resistant glazings.

- Plastics are lighter weight than an equal thickness of glass, which reduces structural requirements for frames.

- Plastics can be formulated to screen out virtually all ultraviolet radiation (UV); glass is not UV absorbing, although coatings and laminates can provide this function.

- Glass is nonflammable but may break in a fire. Fire-rated glass is available. Plastics are typically flammable, may also melt in the heat of a fire, and may give off toxic fumes when they burn.

## IMPROVED GLAZING PRODUCTS

There are three fundamental approaches to improving the energy performance of glazing products:

1. Alter the glazing material itself by changing its chemical composition or physical characteristics. An example of this is tinted glazing. The glazing material can also be altered by creating a laminated glazing.

2. Apply a coating to the glazing material surface. Reflective coatings and films were developed to reduce heat gain and glare, and more recently, low-emittance and spectrally selective coatings have been developed to improve both heating and cooling season performance.

3. Assemble various layers of glazing and control the properties of the spaces between the layers. These strategies include the use of two or more panes or films, low-conductance gas fills between the layers, and thermally improved edge spacers.

Two or more of these approaches may be combined. Each of these improvements to the glazing is discussed below. Thermal improvements to the window sash and frame are discussed in Chapter 4.

### Tinted Glazing

Both plastic and glass materials are available in a large number of tints. The tints absorb a portion of the light and solar heat. Tinting changes the color of the window and can increase visual privacy. The primary uses for tinting are to reduce glare from the bright outdoors and reduce the amount of solar energy transmitted through the glass.

Tinted glazings retain their transparency from the inside, so that the outward view is unobstructed. The most common colors are neutral gray, bronze, and blue-green, which do not greatly alter the perceived color of the view and tend to blend well with other architectural colors. Many other specialty colors are available for particular aesthetic purposes.

Tinted glass is made by altering the chemical formulation of the glass with special additives. Its color changes with the thickness of the glass and the addition of coatings applied after manufacture. Every change in color or combination of differ-

---

**Improved Glazing Products**

- Tinted glazing

- Reflective coatings and films

- Double glazing

- Multiple panes or films

- Low-emittance and spectrally selective coatings

- Gas fills

- Thermally improved edge spacers

- Superwindows

---

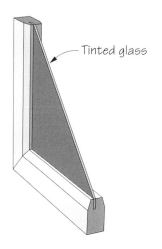

Figure 3-8. Tinted glazing reduces solar heat gain and glare.

ent glass types affects transmittance, solar heat gain coefficient, reflectivity, and other properties. Glass manufacturers list these properties for every color, thickness, and assembly of glass type they produce.

Tinted glazings are specially formulated to maximize their absorption across some or all of the solar spectrum and are often referred to as "heat-absorbing." All of the absorbed solar energy is initially transformed into heat within the glass, thus raising the glass temperature. Depending upon climatic conditions, up to 50 percent of the heat absorbed in a single layer of tinted glass may then be transferred via radiation and convection to the inside. Thus, there may be only a modest reduction in overall solar heat gain compared to other glazings. Heat-absorbing glass provides more effective sun control when used as the outer layer of a double-pane window, as discussed later in the chapter.

There are two categories of tinted glazing: the traditional tints that diminish light as well as heat gain, and spectrally selective tints that reduce heat gain but allow more light to be transmitted to the interior. The traditional tinted glazing often forces a trade-off between visible light and solar gain. For these bronze and gray tints, there is a greater reduction in visible light transmittance than there is in solar heat gain coefficient. This can reduce glare by reducing the apparent brightness of the glass surface, but it also reduces the amount of daylight entering the room. For windows where daylighting is desirable, it may be more satisfactory to use a spectrally selective tint or coating along with other means of controlling solar gain. Tinted glazings can provide a measure of visual privacy during the day when they reduce visibility from the outdoors. However, at night the effect is reversed and it is more difficult to see outdoors from the inside.

To address the problem of reducing daylight with traditional tinted glazing, glass manufacturers have developed new types of tinted glass that are "spectrally selective." They preferentially transmit the daylight portion of the solar spectrum but absorb the near-infrared part of sunlight. This is accomplished by adding special chemicals to the float glass process. Like other tinted glass, they are durable and can be used in both monolithic and multiple-glazed window applications. These glazings have a light blue or green tint and have visible transmittance values higher than conventional bronze- or gray-tinted glass, but have lower solar heat gain coeffi-

> **Characteristics of Tinted Glazing**
>
> - Reduces solar heat gain.
>
> - Lowers visible light transmittance, which reduces glare but diminishes available daylight.
>
> - Alters glass color but views in daytime are not significantly diminished.
>
> **Applications of Tinted Glazing**
>
> - Predominantly in commercial buildings.
>
> - Warm climates where minimizing solar heat gain is a priority.
>
> - Situations where reduced glare is desirable.

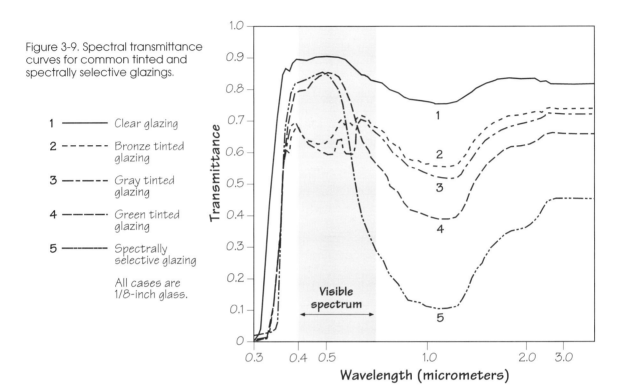

Figure 3-9. Spectral transmittance curves for common tinted and spectrally selective glazings.

1 ———— Clear glazing

2 - - - - - Bronze tinted glazing

3 —— - —— Gray tinted glazing

4 ——— - ——— Green tinted glazing

5 ———————— Spectrally selective glazing

All cases are 1/8-inch glass.

cients. Typical spectral transmittance plots are shown in Figure 3-9. Because they are absorptive, they are best used as the outside glazing in a double-glazed unit, as explained later in this chapter. They can also be combined with low-E coatings to enhance their performance further. Spectrally selective tinted glazings provide a substantial improvement over conventional clear, bronze, and gray glass, and a modest improvement over the existing green and blue-green colored tinted glasses that already have some selectivity.

The visible light transmittance (VT) of traditional tinted glazings is usually reduced substantially more than the solar heat gain coefficient. A ratio between the two factors, termed the light-to-solar gain ratio (LSG), has been developed in order to gauge the relative efficiency of different glass types in transmitting daylight while blocking solar gain (Figure 3-10). The LSG ratio is defined as a ratio between visible transmittance (VT) and solar heat gain coefficient (SHGC). For example, clear glass has a visible light transmittance (VT) of 0.90 and a solar heat gain coefficient (SHGC) of 0.86, so LSG is VT/SHGC

= 1.05. Bronze-tinted glass has a visible transmittance of 0.68 and a solar heat gain coefficient of 0.71, which results in an LSG ratio of 0.96. This illustrates that while the bronze tint lowers the SHGC, it lowers the VT even more. The spectrally selective blue tint has a higher VT of 0.77 and a lower SHGC of 0.58, resulting in an LSG of 1.33—significantly better than the bronze tint. Figure 3-10 shows similar relationships in comparing tinted double glazings, as well as units with a combination of tinted glass and low-E coatings. Glazings with low-E and spectrally selective coatings (discussed later in this chapter) can reduce the SHGC significantly but retain a relatively high VT, producing LSG ratios that are superior to those for tinted glass alone.

Tinted glazing is much more common in commercial windows than in residential windows. Residential solar gains are more typically controlled using blinds, drapes, or other decorative window treatments. However, tinted glazings can be used in combination with the more traditional residential sun controls, where increased solar gain control or privacy is

Figure 3-10. Light-to-solar-gain ratio (LSG) of various types of glass.

| Glass Description | Center-of-Glass Values | | |
| --- | --- | --- | --- |
| | Visible Transmittance (VT) | Solar Heat Gain Coefficient (SHGC) | Light-to-Solar-Gain Ratio (LSG) |
| Single glass | | | |
| Clear | 0.90 | 0.86 | 1.05 |
| Bronze | 0.68 | 0.71 | 0.96 |
| Spectrally selective blue | 0.77 | 0.58 | 1.33 |
| Double glass (inner layer is clear) | | | |
| Clear | 0.82 | 0.75 | 1.20 |
| Bronze | 0.62 | 0.60 | 1.03 |
| Spectrally selective blue | 0.70 | 0.46 | 1.52 |
| Double glass (inner layer is clear with low-E coating) | | | |
| Clear | 0.78 | 0.63 | 1.24 |
| Bronze | 0.58 | 0.49 | 1.18 |
| Spectrally selective blue | 0.66 | 0.35 | 1.84 |

All values in this table are measured at the center of the glass. Total window values for SHGC and visible transmittance will be lower because of the effects of the frame.

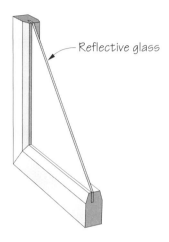

Figure 3-11. Reflective coatings and films reduce solar heat gain.

---

### Characteristics of Reflective Coatings and Films

- Reduce solar heat gain.

- Lower visible light transmittance, which reduces glare but diminishes available daylight.

- May produce exterior mirroring effects.

### Applications of Reflective Coatings and Films

- Predominantly in commercial buildings.

- Warm climates where minimizing solar heat gain is a priority.

- Situations where reduced glare is desirable.

---

needed. In retrofit situations, when windows are not being replaced, tinted plastic film may be applied to the inside surface of the glazing.

Tinted glazing is in place for the life of the window; this has both advantages and disadvantages. On the positive side, the tinted glazing requires no maintenance or operation from the inhabitants. On the other hand, it cannot respond to changing conditions. It cuts out heat and light equally well in the summer and in the winter, when they might be needed.

## Reflective Coatings and Films

As the SHGC is lowered in single-pane tinted glazings, the daylight transmission (VT) drops even faster, and there are practical limits on how low the SHGC can be made using tints. If larger reductions are desired, a reflective coating can be used to lower the solar heat gain coefficient by increasing the surface reflectivity of the material. These coatings usually consist of thin metallic layers. The reflective coatings come in various metallic colors (silver, gold, bronze), and they can be applied to clear or tinted glazing. The solar heat gain coefficient of the substrate can be reduced a little or a lot, depending on the thickness and reflectivity of the coating, and its location on the glass.

Similar to tinted films in retrofit situations, reflective coatings may be applied to the inner glass surface of an existing window by means of an adhesive-bonded, metallic-coated plastic film. The applied films are effective at reducing solar gains but are not as durable as coated glass.

As with tinted glazing, the visible light transmittances of reflective glazings are usually reduced substantially more than the solar heat gain coefficient. Reflective glazings are usually used in commercial buildings for large windows, for hot climates, or for windows where substantial solar heat gains and/or glare are present. In residences, they are usually reserved for special cases. For example, a picture window looking west over a large body of water experiences substantial solar gains and reflected glare during summer afternoons and may require a reflective glazing.

Special thought should always be given to the effect of the reflective glazing on the outside. Acting like a mirror, the reflective glass intensifies the sun's effects, and could momentarily blind pedestrians or drivers, burn plants, or overheat a patio. It is also important to remember that reflective glass acts like a mirror on the side facing the light. Thus, a reflective window that acts like a mirror to the outside during the day will look like a mirror on the inside during the night.

## Double Glazing

As noted in Chapter 1, storm windows added onto the outside of window frames during the stormy winter season were the first double-glazed windows. The intent was to reduce infiltration from winter winds by providing a seal around all the operating sash. Improving the insulating value of the glazing was an important secondary effect.

When manufacturers began to experiment with factory-sealed, double-pane glass to be installed for year-round use, they encountered a number of technical concerns, such as how to allow for different thermal movement between the two panes, how to prevent moisture from forming between the panes and condensing on an inaccessible surface, and how to allow for changes in atmospheric pressure as the assembly was moved from factory to installation site. These issues have been successfully addressed over the years with a variety of manufacturing techniques and material selections.

When double-glass units first came on the market, the two glass layers were often fused around the perimeter to make a permanently sealed air space. In recent years, however, spacers and polymer sealants have largely replaced glass-to-glass seals, and have proven sufficiently durable for residential applications. The layers of glass are separated by and adhere to a spacer, and the sealant, which forms a gas and moisture barrier, is applied around the entire perimeter. Normally, the spacer contains a desiccant material to absorb any residual moisture that may remain in the air space after manufacture. Sealed insulating glass units are now a mature, well proven technology. Designs utilizing high-quality sealants and manufactured with good quality control should last for decades without seal failure.

In addition to sealed insulating glass units, some manufacturers offer double-glazed units with nonsealed removable glazing panels. A blind or shade may be located between the glazings, or the inner glazing may be added in winter and removed in summer. In these double-glazed designs, the inner glazing fits snugly to the sash, and the unsealed air space

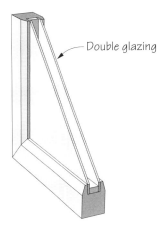

Double glazing

Figure 3-12. Double glazing improves thermal resistance.

### Characteristics of Double Glazing

- Thermal resistance is increased, which reduces winter heat loss and summer heat gain.

- Visible light transmittance is only slightly diminished.

- Best thermal performance occurs with about 1/2-inch (12 mm) space between panes when filled with air.

### Applications of Double Glazing

- All residential buildings with a significant heating season.

- Most residential buildings with a significant cooling season that use frequent or continuous air-conditioning.

Figure 3-13. Impact of multiple glazing on annual energy performance.

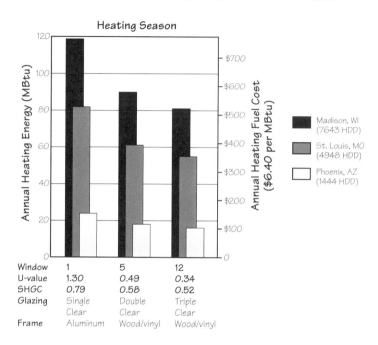

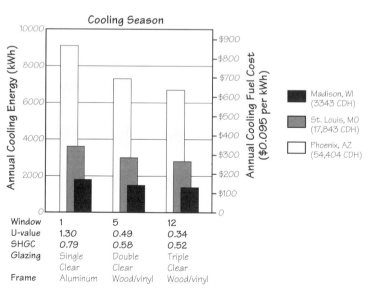

Note: The annual energy performance figures shown here are for a typical 1540 sq ft house. U-factor and SHGC are for total window including frame. House and windows are described in Appendix A. MBtu=millions of Btu. HDD=heating degree days. kWh=kilowatt hours. CDH=cooling degree hours.

is normally vented to the exterior with a small tube to prevent condensation from forming in the air space. Three or more glazing layers can provide even more insulating value.

## Impact on Annual Energy Use

Figure 3-13 illustrates the impact of double and triple glazing on annual energy use in three climates. Even though the total glazing area is a small percentage of the total building envelope, increasing the number of glazing layers has a significant impact on the overall house energy use.

## Glass Coatings and Tints in Double Glazing

Both reflective coatings and tints on double-glazed windows are effective in reducing summer heat gain; however, only certain coatings contribute to reducing winter heat loss, and tints do not affect the U-factor at all. It is possible to provide reflective coatings on any one of the four surfaces, although they are usually located on the outermost surface or on the surfaces facing the air space. Coating location can also depend on the type of coating. Some vacuum-deposited reflective coatings must be placed in a sealed air space because they would not survive exposure to outdoor elements, finger prints, or cleaning agents. Pyrolytic coatings that are created with a high-temperature process as the glass is formed are extremely hard and durable and can be placed anywhere. Each location produces a different visual and heat transfer effect. Other advanced coatings such as low-emittance and spectrally selective coatings are normally applied to double-glazed or triple-glazed windows. These applications are discussed later in this chapter.

Double-pane units can be assembled using different glass types for the inner and outer layers (Figure 3-14). Typically, the inner layer is standard clear glass, while the outer layer can be tinted, reflective, or both. The solar heat gain coefficient is reduced because the tinted glass and clear glass both reduce transmitted radiation. In addition, this design further reduces solar heat gain because the inner clear glass, the gas fill, and any low-E coating keep much of the heat absorbed by the outer glass from entering the building interior.

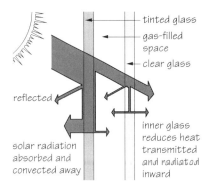

Figure 3-14. Using a tinted outer pane and a clear inner pane in a double-glazed window reduces heat gain. Most heat absorbed by the outer layer does not enter the building interior.

## Gap Width in Multiple Glazed Units

Window manufacturers have some flexibility to reduce heat transfer by selecting the best gap width between two or more glazings. The air space between two pieces of glass reaches its optimum insulating value at about 1/2-inch (12 mm) thickness when filled with air or argon as shown in Figure 3-15. As the gap gets larger, convection increases and slowly increases heat transfer. Below 3/8 inch (9 mm), conduction through the air gap increases and the U-factor rises more rapidly. Krypton gas has its optimum thickness at about 1/4 inch (6 mm), so that if smaller air gaps are required, for example in a three-layer window whose overall exterior dimensions are limited, krypton may be the best selection, although it is also more costly. Figure 3-15 also shows the center-of-glass U-factors for double-glazed units with a low-E coating. Although all of the U-factors are substantially reduced by the presence of the low-E coating, the effects of gap dimension and gas fill are similar to the case of conventional double glazing without coatings.

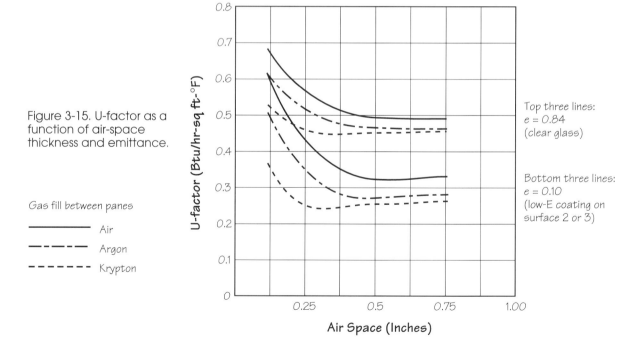

Figure 3-15. U-factor as a function of air-space thickness and emittance.

Gas fill between panes

———————— Air

— — — — — Argon

- - - - - - - Krypton

Top three lines: e = 0.84 (clear glass)

Bottom three lines: e = 0.10 (low-E coating on surface 2 or 3)

## Divided Lights

Manufacturers have been struggling with the problem of many homeowners' preference for traditional, divided light windows, which have many small panes separated by thin bars called muntins. With single-pane glass, true divided lights actually improved the thermal performance of the window because the wood muntins had a higher insulating value than the glass. Some manufacturers have introduced "true divided light" insulated units, in which traditional-looking muntins hold small, individual, insulated panes (Figure 3-16). However, these are expensive and difficult to fabricate with insulated glass and have greater thermal losses due to the number of edges, which now have metal in them.

A second option is to produce a single, large sealed glass unit with "muntins" glued to the inside and outside surfaces,

Figure 3-16. True divided light windows can be created with individual units of insulated glazing between muntins.

Figure 3-17. A realistic appearance of divided lights can be achieved by placing muntins over both sides of a single insulated glazing unit and a grid between the glazings. A lower-cost alternative is to use the grid alone without the interior and exterior muntins.

Figure 3-18. Divided lights are simulated in a simple way by placing muntins over the inside of a single insulated glazing unit.

while a grid is placed in the middle of one large insulated unit, giving the visual effect of divided lights (Figure 3-17). This reduces fabrication costs but does not reduce resistance to heat flow if the muntins in the middle are metal and if they touch both lights of glass.

A third option, which is more energy efficient, is to build a large-pane insulated unit that has snap-on or glued-on grilles to simulate the traditional lights, although these do not always appear authentic (Figure 3-18).

The energy performance of the simple snap-on grid will be similar to a unit without any mullions; however, the true divided lights will result in greater heat transfer because of the additional edges. The energy performance of a unit with a grid set between glazings (Figure 3-17) can be comparable to a unit without any muntins if the grid set between the glazings is at least 1/8 inch (3 mm) away from both lights of glass.

### Special Products

Glass blocks present a very special case of double glazing. They provide light with some degree of visual privacy. Their U-factor is about 0.6; larger blocks are slightly more efficient, as there are proportionately fewer glass edges that conduct heat more readily. Plastic blocks, which have a lower U-factor than glass, are also available. However, when installed, the necessary grouting reduces the energy efficiency. Also, metal mesh and steel reinforcing bars, used between blocks to provide structural stability, provide thermal bridges which also reduce energy efficiency.

Plastic glazings are available in a number of configurations with double layers. Double-glazed acrylic bubble skylights are formed with two layers separated by an air space of varying thickness, ranging from no separation at the edges to as much as 3 inches (7.6 cm) at the top of the bubble. The average separation is used to calculate the effective U-factor.

Multicell polycarbonate sheets, which can be mounted with the divider webs running vertically or horizontally, are available. The divider webs increase the effective insulating value of the glazing by reducing convection exchange within the cells, especially when they are mounted horizontally.

## Multiple Panes or Films

By adding a second pane, the insulating value of the window glass alone is doubled (the U-factor is reduced by half). As expected, adding a third or fourth pane of glass further increases the insulating value of the window, but with diminishing effect.

Triple- and quadruple-glazed windows became commercially available in the 1980s as a response to the desire for more energy-efficient windows. There is a trade-off with this approach, however. As each additional layer of glass adds to the insulating value of the assembly, it also reduces the visible light transmission and the solar heat gain coefficient, thereby reducing the window's value for providing solar gains or daylighting. In addition, other complications are encountered. Additional panes of glass increase the weight of the unit, which makes mounting and handling more difficult and transportation more expensive.

Because of the difficulties discussed above, it is apparent there are physical and economic limits to the number of layers of glass that can be added to a window assembly. However, multiple-pane units are not limited to assemblies of glass. One popular innovation is based on substituting an inner plastic

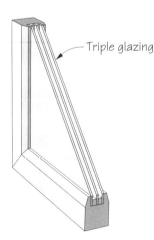

Figure 3-19. Multiple panes further improve thermal resistance.

Triple glazing

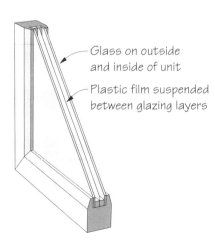

Glass on outside and inside of unit

Plastic film suspended between glazing layers

Figure 3-20. Glazing section with inner layer of thin plastic film.

### Characteristics of Multiple Panes or Films

- Thermal resistance is further increased (more than for double glazing), which reduces winter heat loss and summer heat gain.

- Higher temperatures on the interior glass surface contribute to greater comfort and less condensation in winter.

- Visible light transmittance decreases with each additional layer.

- Solar heat gain coefficient decreases with each additional layer, which reduces cooling loads in summer but reduces beneficial gains in winter as well.

- Additional glass panes increase weight, but plastic films achieve the same purpose without adding much weight.

### Applications of Multiple Panes or Films

- Very cold climates where reduction in heat loss is a major priority.

- Very hot climates.

film for the middle layer of glass. The plastic film is very lightweight, and because it is very thin, it does not increase the thickness of the unit. The glass layers protect the inner layer of plastic from scratching, mechanical abuse, corrosion, weathering, and visual distortions caused by wind pressure. Thus, the strength and durability of plastic as a glazing material are no longer issues when the plastic is protected from physical abuse and weathering by inner and outer layers of glass. The plastic films are specially treated to resist UV degradation and they are heat shrunk so they remain flat under all conditions.

The plastic inner layer serves a number of important functions. It decreases the U-factor of the window assembly by dividing the inner air space into multiple chambers. Units are offered with one or two inner layers of plastic. Secondly, a low-E coating can be placed on the plastic film itself to further lower the U-factor of the assembly. Also, the plastic film can be provided with spectrally selective coatings to reduce solar gain in hot climates without significant loss of visible transmittance. The performance of multiple-pane window assemblies with low-emittance coatings and gas fills is described in the following sections.

## Low-Emittance and Spectrally Selective Coatings

The principal mechanism of heat transfer in multilayer glazing is thermal radiation from a warm pane of glass to a cooler pane. Coating a glass surface with a low-emittance material and facing that coating into the gap between the glass layers blocks a significant amount of this radiant heat transfer, thus lowering the total heat flow through the window. The improvement in insulating value due to the low-E coating is roughly equivalent to adding another pane of glass to a multipane unit.

The solar spectral reflectances of low-E coatings can be manipulated to include specific parts of the visible and infrared spectrum. This is the origin of the term "spectrally selective glazings," which can allow specific portions of the energy spectrum to be "selected," so that desirable wavelengths of energy are transmitted and others specifically reflected. A glazing material can then be designed to optimize energy flows for solar heating, daylighting, and cooling (Figure 3-22).

With conventional clear glazing, a significant amount of solar radiation passes through the window, and then heat

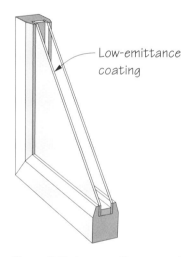

Low-emittance coating

Figure 3-21. Low-emittance and spectrally selective coatings are placed on glass surfaces to control heat loss and heat gain by radiant energy.

from objects within the house is reradiated back through the window (Figure 3-23). For example, a glazing design for maximizing solar gains in the winter would ideally allow all of the solar spectrum to pass through, but would block the reradiation of heat from the inside of the house. The first low-E coatings were designed to have a high solar heat gain coefficient and a high visible transmittance to transmit the maximum amount of sunlight into the interior while reducing the U-factor significantly (Figure 3-24).

A glazing designed to minimize summer heat gains but allow for some daylighting would allow most visible light through, but would block all other portions of the solar spectrum, including ultraviolet light and near-infrared, as well as long-wave heat radiated from outside objects, such as paving and adjacent buildings. These second-generation low-E coatings were designed to reflect the solar near-infrared, thus reducing the total solar heat gain coefficient while maintaining high levels of light transmission (Figure 3-25). Variations on this design (modified coatings and/or glazings) can further reduce summer solar heat gain and control glare.

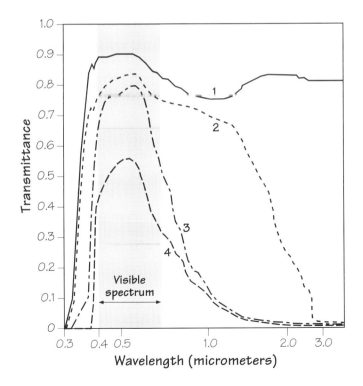

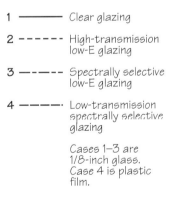

Figure 3-22. Spectral transmittance curves for glazings with low-emittance and spectrally selective coatings.

1 ———— Clear glazing

2 -  -  -  - High-transmission low-E glazing

3 —  ·  —  · Spectrally selective low-E glazing

4 —  —  —  · Low-transmission spectrally selective glazing

Cases 1–3 are 1/8-inch glass. Case 4 is plastic film.

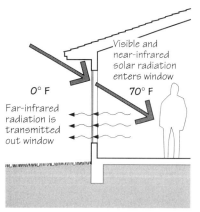

Figure 3-23: Clear glass allows solar heat gain and but does not reduce winter heat loss.

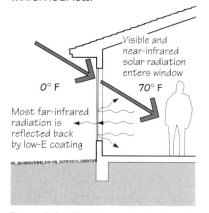

Figure 3-24: High-transmission low-E glass provides solar heat gain and reduces winter heat loss.

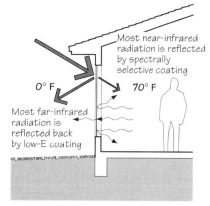

Figure 3-25: Selective-transmission low-E glass reduces winter heat loss and summer heat gain.

There are three basic types of low-E coatings available on the market today:

1. High-transmission low-E

   High transmission products are best suited to buildings located in heating-dominated climates and particularly to south-facing windows in passive solar designs (Figure 3-24).

2. Selective-transmission low-E

   Selective-transmission products are suited for buildings that have both winter heating and summer cooling requirements. The low-emittance characteristics of this glass reduce winter heat loss. In the summertime, the selective properties allow natural daylighting, but block a large fraction of the solar infrared energy, thus reducing cooling loads (Figure 3-25). A similar effect can be achieved by combining a high-transmittance, high SHGC, low-E coating with a spectrally selective tinted glass.

3. Low-transmission low-E

   Placing a low-E coating on dark tinted glass, and/or increasing the solar reflectance of the coating itself, creates a product with the insulating capability of low-E coatings, along with glare control and a high level of solar heat rejection. This is especially suited to controlling solar gain and glare in cooling-dominated climates.

Window manufacturers' product information may not list emittance ratings. Rather, the effect of the low-E coating is incorporated into the U-factor for the unit or glazing assembly. The type and quality of low-E coating will affect not only the U-factor, but also the transmittance and solar heat gain coefficient of a glass. All these properties (U-factor, VT, and SHGC) need to be taken into consideration in selecting a particular glazing product.

## Energy Performance Impacts of Coatings

Figure 3-26 illustrates the impact of various low-E coatings on the windows of a typical house in three different climates. The U-factors of the three coatings (windows 7, 9, and 10) are approximately equal, making winter heat losses similar. The variation in SHGC, however, influences both the heating and

Figure 3-26. Impact of low-emittance and spectrally selective coatings on whole house annual energy performance.

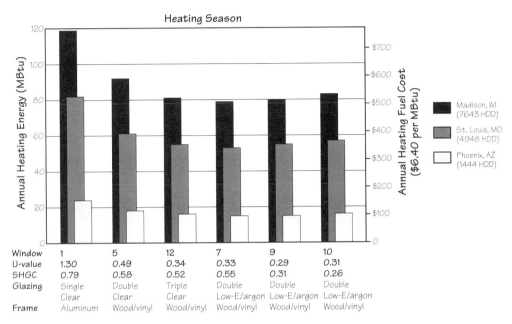

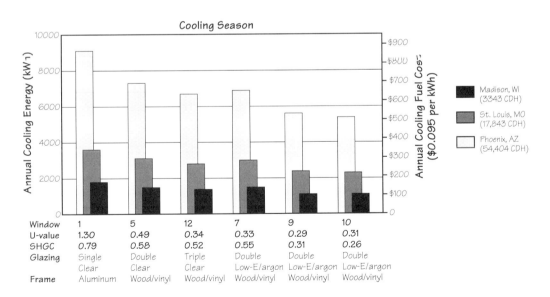

Note: The annual energy performance figures shown here are for a typical 1540 sq ft house. U-factor and SHGC are for total window including frame. House and windows are described in Appendix A. MBtu=millions of Btu. HDD=heating degree days. kWh=kilowatt hours. CDH=cooling degree hours.

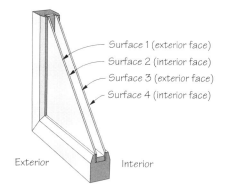

Figure 3-27. Low-E coatings are placed on different surfaces to achieve different effects. The placement of the coating does not affect the U-factor. A coating placed on the outside of the inner glazing surface (#3) is most effective for maximizing passive solar gain. A coating placed on the inside of the outer glazing surface (#2) is most effective for minimizing solar heat gain.

cooling season energy use. The lower SHGC in windows 9 and 10 makes a notable improvement in cooling energy use, but actually increases heating season energy use because of the reduced passive solar gain in winter. Note that a double-glazed window with a high-transmission low-E coating (window 7 in Figure 3-26) has about the same heating season performance as a clear triple-glazed window (window 12). To determine which glazing would be best on an annual basis, the specific costs of heating and cooling energy for a given location would have to be accounted for.

## Coating Placement

The placement of a low-E coating within the air gap of a double-glazed window does not affect the U-factor but it does influence the solar heat gain coefficient (SHGC). That is why, in heating-dominated climates, placing a low-E coating on the #3 surface (outside surface of the inner pane) is recommended to maximize winter passive solar gain at the expense of a slight reduction in the ability to control summer heat gain (Figure 3-27). In cooling climates, a coating on the #2 surface (inside surface of the outer pane) is generally best to reduce solar heat gain and maximize energy efficiency. Manufacturers sometimes place the coatings on other surfaces (e.g., #2 surface in a heating climate) for other reasons, such as minimizing the potential for thermal stress. Multiple low-E coatings are also placed on surfaces within a triple-glazed window assembly, or on the inner plastic glazing layers of multipane assemblies referred to as superwindows (discussed later in this chapter), with a cumulative effect of further improving the overall U-factor.

## Coating Types

There are two basic types of low-E coatings—sputtered and pyrolytic, referring to the process by which they are made. The best of each type of coating is colorless and optically clear. Some coatings may have a slight hue or subtle reflective quality, particularly when viewed in certain lighting conditions or at oblique angles.

A sputtered coating is multilayered (typically, three primary layers, with at least one layer of metal) and is deposited on glass or plastic film in a vacuum chamber. The total thickness of a sputtered coating is only 1/10,000 of the thickness of a human hair. Sputtered coatings often use a silver layer and must be protected from humidity and contact. For this reason they are sometimes referred to as "soft coats." Since sputtering is a low-temperature process, these coatings can be deposited on flat sheets of glass or thin plastic films. While sputtered coatings are not durable in themselves, when placed into a sealed double- or triple-glazed assembly they should last as long as the sealed glass unit. Sputtered coatings typically have lower emittances than pyrolytic coatings. They are available commercially with emittance ratings of $e = 0.20$ to as low as $e = 0.04$ ($e = 0.20$ means that 80 percent of the long-wavelength radiant energy received by the surface is reflected, while $e = 0.04$ means 96 percent is reflected). For uncoated glass, $e = 0.84$, which means only 16 percent of the radiant energy received by the surface is reflected.

A typical pyrolytic coating is a metallic oxide, most commonly tin oxide with some additives, which is deposited directly onto a glass surface while it is still hot. The result is a baked-on surface layer that is quite hard and thus very durable, which is why this is sometimes referred to as a "hard coat." A pyrolytic coating can be ten to twenty times thicker than a sputtered coating but is still extremely thin. Pyrolytic coatings can be exposed to air, cleaned with normal cleaning products, and subjected to general wear and tear without losing their low-E properties.

Because of their greater durability, pyrolytic coatings are available on single-pane glass and separate storm windows, but not on plastics, since they require a high-temperature process. In general, though, pyrolytic coatings are used in sealed, double-glazed units with the low-E surface inside the sealed air space. While there is considerable variation in the specific properties of these coatings, they typically have emittance ratings in the range of $e = 0.40$ to $e = 0.15$.

Spectrally selective coatings are modified versions of sputtered low-E coatings. The number of layers and their thicknesses are altered, which causes the coating to reflect the sun's near-infrared energy as well as the long-wave infrared. Pyrolytic coatings that are strongly spectrally selective are not readily available now, although they are under development. However, by adding a clear low-E pyrolytic coating to a spectrally selective tinted glazing, a similar effect can be achieved.

---

**Characteristics of Low-E and Selective Coatings**

- Coatings can be formulated to reflect long-wave radiant heat, giving an improved U-factor and reducing winter heat loss.

- Higher temperatures on the interior glass surface contribute to greater comfort and less condensation in winter.

- Coatings can be formulated to reflect solar radiation back toward the exterior, resulting in reduced summer heat gain.

- Coatings can be formulated so that visible light transmittance is only slightly affected.

**Application of Low-E Coatings**

- Cold climates where reducing heat loss is a priority.

**Application of Selective Coatings**

- Warm climates where minimizing solar heat gain is a priority.

A laminated glass with a spectrally selective low-E sputtered coating on plastic film sandwiched between two layers of glass offers the energy performance of single-pane, spectrally selective glass and the safety protection of laminated glass. However, in this configuration, since the low-E surface is not exposed to an air space, there is no effect on the glazing U-factor.

Spectrally selective coatings on plastic can also be applied to existing glass as a retrofit measure, thus reducing the SHGC of an existing clear glass considerably while maintaining a high visible light transmittance. Other conventional tinted and reflective films will also reduce the SHGC but at the cost of lower visible transmittance. Reflective films can also lower the U-factor, since the surface facing the room has a lower emittance than uncoated glass.

The technology to produce low-E and selective coatings has been evolving quickly, and the market price has been dropping at a corresponding rate. Once the initial hurdle of making a substantial investment in the necessary production machinery has been met, there is a market incentive for manufacturers to provide coatings on as much of the glass produced as possible. Thus, windows with low-E coatings have become a standard residential window product. Approximately 40 percent of window products now sold in the United States have low-E coatings, and the figure will continue to rise.

## Gas Fills

Another improvement that can be made to the thermal performance of insulating glazing units is to reduce the conductance of the air space between the layers. Originally, the space was filled with air or flushed with dry nitrogen just prior to sealing. In a sealed glass insulating unit air currents between the two panes of glazing carry heat to the top of the unit and settle into cold pools at the bottom. Filling the space with a less conductive, more viscous, or slow-moving gas minimizes the convection currents within the space, conduction through the gas is reduced, and the overall transfer of heat between the inside and outside is reduced.

Manufacturers have introduced the use of argon and krypton gas fills, with measurable improvement in thermal performance (see Figure 3-15). Argon is inexpensive, non-

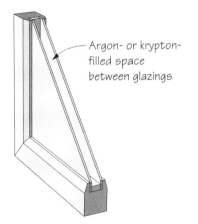

Argon- or krypton-filled space between glazings

Figure 3-28. Argon and krypton gas fills increase thermal resistance.

toxic, nonreactive, clear, and odorless. The optimal spacing for an argon-filled unit is the same as for air, about 1/2 inch (12 mm). Krypton has better thermal performance, but is more expensive to produce. Krypton is particularly useful when the space between glazings must be thinner than normally desired, for example, 1/4 inch (6 mm). A mixture of krypton and argon gases is also used as a compromise between thermal performance and cost.

Filling the sealed unit completely with argon or krypton presents challenges that manufacturers continue to work on. A typical gas fill system adds the gas into the cavity with a pipe inserted through a hole at the edge of the unit. As the gas is pumped in, it mixes with the air, making it difficult to achieve 100 percent purity. Recent research indicates that 90 percent is the typical concentration achieved by manufacturers today. Some manufacturers are able to consistently achieve better than 95 percent gas fill by using a vacuum chamber. An uncoated double-pane unit filled with 90 percent argon gas and 10 percent air yields a slightly more than 5 percent improvement in the insulating value at the center of the glass, compared to the same unit filled with air. However, when argon and krypton fills are combined with low-E coatings and multipane glazings, more significant reductions of 15 to 20 percent can be achieved. Since the low-E coating has substantially reduced the radiation component of heat loss, the gas fill now has a greater proportional effect on the remaining heat transfer by convection and conduction.

Some people express concern at the idea of a strange gas leaking out from their window units. The gases are inert, nontoxic, and occur naturally in the atmosphere, but maintaining long-term thermal performance is certainly an issue. Studies have shown less than 0.5 percent leakage per year in a well-designed and well-fabricated unit, or a 10 percent loss in total gas over a twenty-year period. The effect of a 10 percent gas loss would only be a few percent change in U-factor on an overall product basis. Keeping the gas within the glazing unit depends largely upon the quality of the design, materials, and, most important, assembly of the glazing unit seals.

### Characteristics of Gas Fills

- Thermal resistance is increased with argon and krypton gas fills, reducing winter heat loss and summer heat gain through conduction.

- Higher temperatures in winter on the interior glass surface contribute to greater comfort and less condensation.

- Visible light transmittance is not affected.

### Application of Gas Fills

- Cold climates where reducing heat loss is a priority.

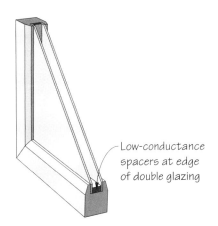

Low-conductance spacers at edge of double glazing

Figure 3-29. Thermally improved spacers reduce heat loss and condensation.

**Characteristics of Thermally Improved Spacers**

- Overall U-factor is improved because heat loss at the glass edges is reduced.

- Higher temperatures on the glass edges produce less condensation.

**Application of Thermally Improved Spacers**

- Cold climates where reducing heat loss is a priority.

## Thermally Improved Edge Spacers

The layers of glazing in an insulating unit must be held apart at the appropriate distance by spacers. The spacer system must provide a number of additional functions in addition to keeping the glass units at the proper dimension:

- accommodate stress induced by thermal expansion and pressure differences;

- provide a moisture barrier that prevents passage of water or water vapor that would fog the unit;

- provide a gas-tight seal that prevents the loss of any special low-conductance gas in the air space;

- create an insulating barrier that reduces the formation of interior condensation at the edge.

Older double-pane wood windows used a wood spacer that could not be hermetically sealed and thus was vented to the outside to reduce fogging in the air gap. Modern versions of this system function well but, because they are not hermetically sealed, cannot be used with special gas fills or some types of low-E coatings. Early glass units were often fabricated with an integral welded glass-to-glass seal. These units did not leak but were difficult and costly to fabricate, and typically had a less-than-optimal narrow spacing. The standard solution for insulating glass units (IGUs) that accompanied the tremendous increase in market share of insulating glass in the 1980s was the use of metal spacers, and sealants. These spacers, typically aluminum, also contain a desiccant that absorbs residual moisture. The spacer is sealed to the two glass layers with organic sealants that both provide structural support and act as a moisture barrier. There are two generic systems for such IGUs: a single-seal spacer and a double-seal system.

In the single-seal system (Figure 3-30), an organic sealant, typically a butyl material, is applied behind the spacer and serves both to hold the unit together and to prevent moisture intrusion. These seals are normally not adequate to contain special low-conductance gases.

In a double-seal system (Figure 3-31), a primary sealant, typically butyl, seals the spacer to the glass to prevent moisture migration and gas loss, and a secondary backing sealant, often silicone, provides structural strength. When sputtered low-E coatings are used with double-seal systems, the coating must be removed from the edge first ("edge deletion") to provide a better edge seal.

Since aluminum is an excellent conductor of heat, the aluminum spacer used in most standard edge systems represented a significant thermal "short circuit" at the edge of the IGU, which reduces the benefits of improved glazings. As the industry has switched from standard double-glazed IGUs to units with low-E coatings and gas fills, the effect of this edge loss becomes even more pronounced. Under winter conditions, the typical aluminum spacer would increase the U-factor of a low-E, gas fill unit slightly more than it would increase the U-factor of a standard double-glazed IGU. The smaller the glass area, the larger the effect of the edge on the overall product properties. In addition to the increased heat loss, the colder edge is more prone to condensation. The temperature effect at the edge can be clearly seen in a series of infrared images shown on page 78 (Figures 3-40 to 3-43).

Window manufacturers have developed a series of innovative edge systems to address these problems, including solutions that depend on material substitutions as well as radically new designs. One approach to reducing heat loss has been to replace the aluminum spacer with a metal that is less conductive, e.g., stainless steel, and change the cross-sectional shape of the spacer (Figures 3-32 and 3-33). These designs are widely used in windows today.

Another approach is to replace the metal with a design that uses materials that are better insulating. The most commonly used design incorporates spacer, sealer, and desiccant in a single tape element. The tape includes a solid, extruded thermoplastic compound that contains a blend of desiccant materials and incorporates a thin, fluted metal shim of aluminum or stainless steel (Figure 3-34). Another approach uses an insulating silicone foam spacer that incorporates a desiccant and has a high-strength adhesive at its edges to bond to glass (Figure 3-35). The foam is backed with a secondary sealant. Both extruded vinyl and pultruded fiberglass spacers have also been used in place of metal designs.

There are several hybrid designs that incorporate thermal breaks in metal spacers or use one or more of the elements described above. Some of these are specifically designed to accommodate three- and four-layer glazings or IGUs incorporating stretched plastic films. All are designed to interrupt the heat transfer pathway at the glazing edge between two or more glazing layers (Figures 3-36 and 3-37).

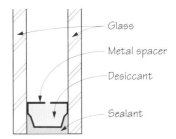

Figure 3-30. Single-seal metal spacer.

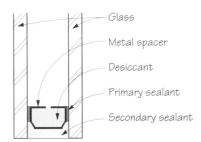

Figure 3-31. Double-seal metal spacer.

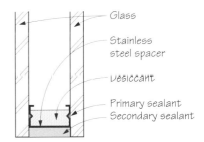

Figure 3-32. Stainless steel spacer.

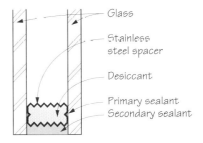

Figure 3-33. Stainless steel spacer.

Warm edge spacers have become increasingly important as manufacturers switch from conventional double glazing to higher-performance glazing. For purposes of determining the overall window U-factor, the edge spacer has an effect that extends beyond the physical size of the spacer to a band about 2-1/2 inches (64 mm) wide. The contribution of this 2-1/2-inch-wide "glass edge" to the total window U-factor depends on the size of the window. Glass edge effects are more important for smaller windows, which have a proportionately larger glass edge area. For a typical residential-size window (3 by 4 feet/0.8 by 1.2 meters), changing from a standard aluminum edge to a good-quality warm edge will reduce the overall window U-factor by .01 to .02 Btu/hr-sq ft-°F.

A more significant benefit may be the rise in interior surface temperature at the bottom edge of the window, which is most subject to condensation. Here, a thermally improved spacer could result in temperature increases of 6–8°F (3–4°C) at the window sightline—or 4–6°F (2–4°C) at a point one inch in from the sightline, which is an important improvement. As new highly insulating "superwindows" are developed, the improved edge spacer becomes an even more important element.

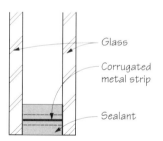

Figure 3-34. Butyl tape spacer with metal.

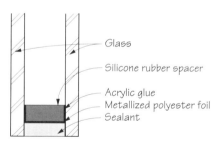

Figure 3-35. Silicone foam spacer.

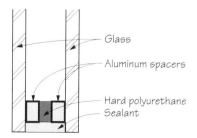

Figure 3-36. Aluminum spacer with thermal break.

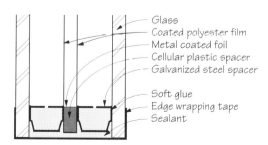

Figure 3-37. Aluminum spacer with thermal break.

## Superwindows

When advanced glazing products are offered on the market, they are generally available as an integrated product, not as isolated features to be selected individually. A window with a combination of all of the measures available to maximize insulating value is referred to as a "superwindow." Superwindows are currently commercially available, but the cost is higher than for the more common double-glazed windows with low-E coatings. Most superwindows developed so far consist of a triple- or quadruple-layer design with two low-E coatings, krypton or argon gas fills, low-conductance spacers, and thermally improved frames. The center glazing layer(s) may be glass or plastic film.

The result is a highly transparent window with a center-of-glass U-factor below 0.15, an overall U-factor below 0.20, and a total SHGC of about 0.40. With continuing improvements in frame and spacer design, overall window U-factors as low as 0.10 seem possible. Such superwindows have a remarkable energy performance. The loss of heat is so low that the diffuse sunlight gained through a north-facing window on a cold, cloudy winter day is greater than the heat losses over the full day. At this threshold of performance, a window can take on a new role in buildings as a net energy provider rather than a net energy loser. Thus, a superwindow has a lower seasonal heating loss than even a highly insulated wall in a cold climate. South-facing windows have always had this property, however, highly efficient superwindow technologies mean that a window facing in any direction can be a passive solar collector.

Figure 3-39 illustrates the impact of various combinations of advanced glazing technologies on annual energy performance for a typical house in three climates. The superwindow configuration modeled here represents an improvement in U-factor but not in SHGC when compared to the double-glazed low-E case. Consequently, the main impact is seen in the heating season, not the cooling season. Note that the double-glazed case with a selective coating (SHGC=0.26) has the best cooling season performance. The superwindow has an added performance advantage not shown directly in these figures—improved comfort. The wintertime inside surface temperatures of the superwindow are much higher than conventional windows during cold weather, thus improving comfort in the house interior, particularly where there are large window areas.

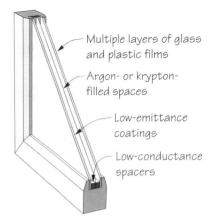

Figure 3-38. Superwindows improve thermal resistance with multiple layers of glass and suspended plastic films, argon and krypton gas fills, low-emittance coatings, and low-conductance spacers.

Multiple layers of glass and plastic films

Argon- or krypton-filled spaces

Low-emittance coatings

Low-conductance spacers

**Characteristics of Superwindows**

- Thermal resistance is increased with multiple layers of glass and suspended plastic films, argon and krypton gas fills, and low emittance coatings. This reduces winter heat loss and summer heat gain.

- Visible light transmittance is diminished slightly with each additional layer.

**Application of Superwindows**

- Cold climates where reduction in heat loss is a priority.

- Where comfort is important, particularly with large glazing areas.

Figure 3-39. Impact of improved glazing technologies on annual energy performance.

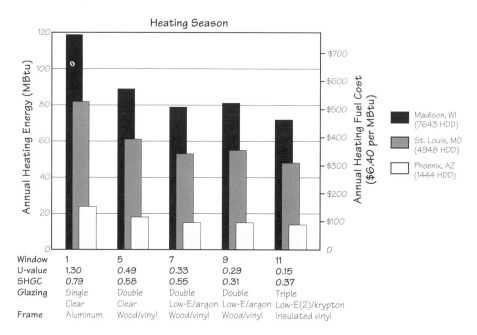

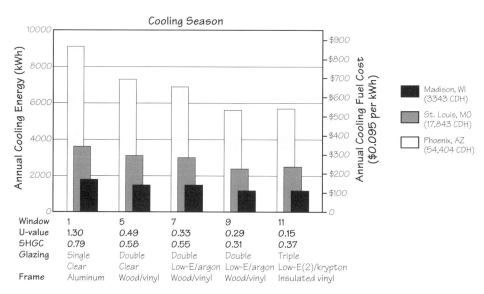

Note: The annual energy performance figures shown here are for a typical 1540 sq ft house. U-factor and SHGC are for total window including frame. House and windows are described in Appendix A. MBtu=millions of Btu. HDD=heating degree days. kWh=kilowatt hours. CDH=cooling degree hours.

# USING THERMOGRAPHY

The human eye is not able to see the heat that radiates from all objects. By using infrared thermography, researchers can "see" and measure the heat that radiates from every surface. This technique can be used to create images, called thermograms, that reflect the temperature of objects. The thermograms on the following pages illustrate a number of thermal effects that occur in window assemblies.

## Comparison of Insulating Glazing Units

Infrared thermography is a testing technique that provides detailed maps of surface temperatures and can visually indicate the specific location of design or materials defects that can cause additional heat loss. The glazings are mounted in a test chamber, the temperature on the outdoor side is lowered, and the special infrared detector in the imager observes the indoor warm side of the glazing, recording its surface temperature, or thermogram. The four images (Figures 3-40 through 3-43) show graphically the effects of warm edge spacer design and high-performance insulating glazings with low-E coatings. These images show the insulating glass unit only—there is no sash or frame.

Figures 3-40 and 3-41 compare an aluminum spacer to a foam spacer, both in a clear double-glazed IGU. The aluminum spacer shows very low temperatures around the entire edge, with the lowest at the bottom as expected, since cold air currents in the IGU settle to the bottom. The unit with a foam spacer shows a similar top-to-bottom trend, but the temperatures at the edge are all substantially higher and the width of the glass edge effect is reduced. Note that the coldest point at the bottom edge of the foam unit (32°F/0°C) is warmer than the warmest upper edge of the unit with the aluminum spacer (20°F/-6.5°C).

A comparison of Figures 3-41 and 3-42 shows the benefit of replacing a clear uncoated glass with a low-E coated glazing in IGUs with identical foam edge spacers. Under these test conditions, the low-E center-of-glass temperature is 8°F (4.3°C) warmer than the uncoated glazing. Both glazings show the expected top-to-bottom temperature gradient, and illustrate visually where condensation would probably occur first if it were cold enough outdoors.

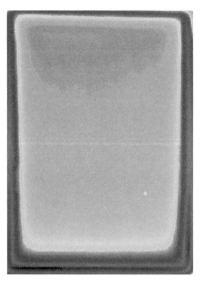

Figure 3-40. Thermogram of a double-glazed window with clear glass and aluminum edge spacers.

Figure 3-41. Thermogram of a double-glazed window with clear glass and foam edge spacers.

20　24　28　32　36　40　44　48　52　56　60　64　68
Temperature (°F)

Figure 3-42. Thermogram of a double-glazed window with a low-E coating and foam edge spacers.

Figure 3-43. Thermogram of superwindow.

Comparing Figures 3-42 and 3-43 shows the added benefit of going to a superwindow glazing, a quadruple-glazed unit with three low-E coatings and krypton gas fill. There is a significant increase in center-of-glass temperature, to 64°F (18°F) from the 53°F (11.8°C) of the low-E unit. Note that the heat loss through the center of the glazing of the superwindow is much lower than at the edge, suggesting that more development work is still needed to improve the edge design.

## Comparison of Vinyl Window Frames

Figures 3-44 and 3-45 show thermograms of two vinyl windows with identical highly insulating glazings. Both frame designs use the same vinyl profile, but one is hollow and the other is filled with foam insulation. Several conclusions can be inferred from detailed study of these images. The glazing performs about the same as expected in both cases. The edge of the glass is much colder than the center, suggesting that these products could benefit from an even better edge spacer.

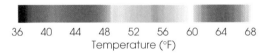

36    40    44    48    52    56    60    64    68
Temperature (°F)

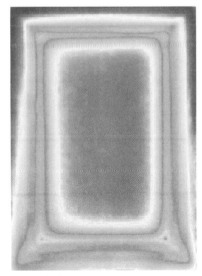

Figure 3-44. Thermogram of a window using a vinyl frame with hollow cores.

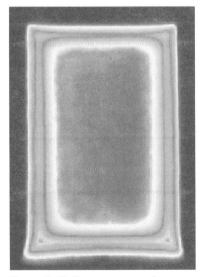

Figure 3-45. Thermogram of a window with an insulated vinyl frame.

The effect of heat conduction from the spacer is also evident in the adjacent frame section in Figure 3-44 with hollow cavities in the vinyl extrusion. Figure 3-44 also shows that air is circulating in the hollow vinyl window frame, and the coldest air is collecting in the cavities at the bottom, representing a significant additional source of heat loss. By comparison, the vinyl frame in Figure 3-45 has been filled with insulation, thus eliminating air flow, and raising the surface temperatures over the entire frame.

## EMERGING AND FUTURE DEVELOPMENTS

The advanced glazing technologies discussed above are on the market already, but there are many other technologies emerging from the research and development pipeline. Some of the promising ones are described below.

Figure 3-46. Thermogram of an evacuated window. Note the pattern of small glass pillars required to separate the two panes of glass.

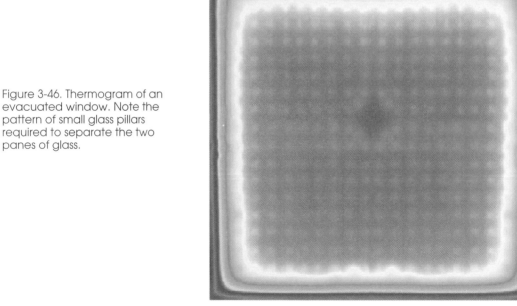

32   36   40   44   48   52   56   60   64   68
Temperature (°F)

## Evacuated Windows

The most thermally efficient gas fill would be no gas at all—a vacuum. A number of researchers around the world have been pursuing the development of insulating window units in which the space between the glazings is evacuated. If the vacuum is low enough, there would be no conductive or convective heat exchange between the panes of glass, thus lowering the U-factor. A vacuum glazing must have a good low-E coating to reduce radiative heat transfer—the vacuum effect alone is not adequate. This principle has been used in the fabrication of thermos bottles for many years, where the silver coating serves as the low-emittance surface.

However, evacuated window assemblies present a number of engineering problems. One major issue is the structural requirement to resist normal air pressure and variable pressures caused by wind and vibration. There can be large thermal stresses between large, window-sized panes of glass. A thermos bottle resists these forces easily because of its strong circular shape. However, the large, flat surfaces of a window tend to bow and flex with changing pressures. Minute glass pillars or spheres have been used in prototypes to maintain the separation between the panes. The pillars are very small but are still somewhat visible, reducing the clarity of the window. Figure 3-46 is a thermographic image of an evacuated window in which the pattern of the pillars is evident. The grid of pillars shows up because of the glass-to-glass conduction through the pillars.

Another issue is the maintenance of an airtight seal around the edge of the unit. The seal must be maintained to eliminate gaseous conduction by keeping the air density within the unit to less than one millionth of normal atmospheric pressure. An air density of only ten times this amount is sufficient to re-establish conduction to its normal value. This vacuum seal must remain intact for the life of the window, through manufacture, transportation, installation, and normal operation and weathering. Special solder glass seals have been used successfully by Australian researchers in the development of large prototypes. Center-of-glass U-factors of 0.2 have been achieved to date, with the possibility of reaching 0.12, while maintaining a high SHGC.

## Aerogel

Aerogel is a silica-based, open-cell, foam-like material composed of about 4 percent silica and 96 percent air. The microscopic cells of the foam entrap gas, thereby preventing convection, while still allowing light to pass. The particles that make up the thin cell walls slightly diffuse the light passing through, creating a bluish haze similar to that of the sky.

Aerogel has received research attention for its ability to be both highly transparent and insulating, making it one of a number of materials that are generically referred to as "transparent insulation." It should be technically possible to produce windows made of aerogel with a center-of-glass U-factor as low as 0.05. However, so far aerogel has only been produced in small quantities and small sizes, so that only tile-sized samples of aerogel have been used as windows for research purposes. European manufacturers have produced double-glazed windows filled with small beads of aerogel. Although the units have good insulating values, they are diffusing and do not provide a view. Aerogels might find a future application as a component of a larger window system, such as spacers between insulating panes of glass, or in skylights or glass blocks.

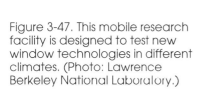

Figure 3-47. This mobile research facility is designed to test new window technologies in different climates. (Photo: Lawrence Berkeley National Laboratory.)

## Smart Windows

"Smart windows" are able to dynamically change their solar-optical properties in response to changing performance requirements. There are two basic types of smart windows—passive devices that respond directly to environmental conditions such as light level or temperature, and active devices that can be directly controlled in response to occupant preferences or heating and cooling system requirements. The main passive devices are photochromics and thermochromics; active devices include liquid crystal, dispersed particle, and electrochromics.

### *Photochromics*

Photochromic materials change their transparency in response to changes in the intensity of light. Photochromic materials have been used in sunglasses that change from clear in the dim indoor light to dark in the bright outdoors. Photochromics may be useful in conjunction with daylighting, allowing just enough light through for lighting purposes, while cutting out excess sunlight that creates glare and overloads the cooling system. Although small units have been produced in volume as a consumer product, cost-effective, large, durable glazings for windows are not yet commercially available.

### *Thermochromics*

Thermochromics change transparency in response to changes in temperature. The materials currently under development are gels sandwiched between glass and plastic that switch from a clear state when cold to a more diffuse, white, reflective state when hot. In their switched-on state, the view through the glazing is lost. Such windows could, in effect, turn off the sunlight when the cooling loads become too high. Thermochromics could be very useful to control overheating for passive solar heating applications. The temperature of the glass, which is a function of solar intensity and outdoor and indoor temperature, would regulate the amount of sunlight reaching the thermal storage element. Thermochromics would be particularly appropriate for skylights because the obscured state would not interfere with views as it would in a typical window. Such units would come with a preset switching temperature, which would have to be selected carefully for the application in mind. Prototype glazings have been tested and may be commercially available in the near future.

## Liquid Crystal Glazing

A variant of the liquid crystal display technology used in wristwatches is now serving as a privacy glazing for new windows. A very thin layer of liquid crystals is sandwiched between two transparent electrical conductors on thin plastic films and the entire package is laminated between two layers of glass. When the power is off, the liquid crystals are in a random and unaligned state. They scatter light and the glass appears as a diffusing layer, which obscures direct view and provides privacy. The material transmits most of the incident sunlight in a diffuse mode; thus its solar heat gain coefficient remains high. When power is applied, the electric field in the device aligns the liquid crystals and the glazing becomes clear in a fraction of a second, permitting view in both directions. The device has only two states, clear and diffusing, and the power must be continuously applied for the glazing to remain in the clear state. This technology is commercially available today for architectural applications. It requires standard house-hold voltage (120v) and is relatively costly but may meet a need where automatic control of privacy or glare from direct sun are needed.

## Particle Dispersed Glazing

This electrically controlled film utilizes a thin liquid-like layer in which numerous microscopic particles are suspended. In its unpowered state the particles are randomly oriented and thus partially block sunlight transmission and view. Transparent electrical conductors allow an electric field to be applied to the dispersed particle film, aligning the particles and raising the transmittance. This glazing system can be partially transmissive between its clear and dark states. However, it is also expensive and is not yet available as a commercial window product.

## Electrochromic Glazing

The most promising smart window technology may be devices based on electrochromic coatings. Although not yet commercially available, they appear to have a good chance to meet performance, cost, and manufacturing requirements that would result in a marketable window system.

Electrochromics can change transparency over a wide continuous range, from about 10 to 70 percent light transmittance in less than one minute, with a corresponding wide range of control over solar heat gain. The coating darkens as it switches and provides a view out under all switching conditions. Switching occurs at very low voltage (1–2 volts), so power supplies and wiring should not be expensive. Removing the voltage stops the electrochromic process without affecting the window's present state of transmittance. Reversing the voltage returns the window to its original state. The reversible process can continue over thousands of cycles.

Electrochromic technology has been actively researched for over fifteen years throughout the world, and promising laboratory results have led to initial prototype development. Examples of electrochromic window prototypes have been demonstrated in a number of buildings in Japan. Electrochromic mirrors are available commercially for rearview mirrors in automobiles and trucks. Electrochromic glazings have also been installed as prototype sunroofs in

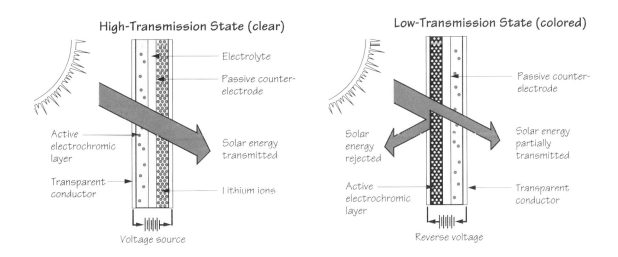

Figure 3-48. Schematic diagram of a five-layer electrochromic coating (not to scale). A reversible low-voltage source moves ions back and forth between an active electrochromic layer and a passive counterelectrode.

cars. Before these sunroof coatings become commercially available, they will have to pass a battery of accelerated durability tests and field tests to prove their long-term performance.

In contrast to thermochromic or photochromic glazings, electrochromic windows offer flexibility in control. Thermochromics and photochromics must have a preset threshold point at which they will change state. Although a range of thresholds could be specified, the specified threshold becomes a permanent feature of that window. Electrochromics, on the other hand, could be operated by computer, by occupant signal, or by simple thermostat or photocell controls. This offers the potential for occupant adjustments and calibration to changing local conditions.

Initial studies show that smart windows are likely to have the greatest energy benefits in commercial buildings in which costs of cooling energy and related mechanical equipment are most significant. By increasing the use of daylighting and reducing the cooling loads, smart windows not only reduce annual energy costs, but also reduce the peak electric demand charges and the required size of the cooling equipment, both of which can be very substantial costs in commercial buildings.

Many of these benefits, however, may also be of value in homes, especially in more extreme climates. In principle, most of the benefits of smart windows might be captured by the intelligent and consistent use of existing manual or automated window control technology. But experience shows that people are often inconsistent and unpredictable in drawing curtains or adjusting blinds, and the cost of motorized controls is an obstacle. By making these control features an integral and dependable part of the building, smart windows can make potential energy savings predictable and guaranteed. The first commercially available products should be on the market by 2000.

# CHAPTER 4

# The Complete Window Assembly

Although glazing materials are the focus of much of the innovation and improvement in windows, the overall performance of any unit is determined by the complete window assembly. The assembly includes the operating and fixed parts of the window frame as well as associated hardware and accessories. These are defined and illustrated at the beginning of this chapter. The next two sections address the different options available for window sash operation and new advances in frame materials designed to improve window energy efficiency. Proper installation of windows is an important aspect of their performance as well. The final section of this chapter discusses other installation issues.

Figure 4 1. New materials and improvements to all parts of the window assembly have contributed to higher- performance windows. These windows with vinyl frames have a high insulating value. (Photo: Summit Window & Patio Door.)

## WINDOW SASH OPERATION

When you select a window, there are numerous operating types to consider. Traditional operable window types include the projected or hinged types such as casement, awning, and hopper, and the sliding types such as double- and single-hung and horizontal sliding (Figure 4-2). In addition, the current window market includes storm windows, sliding and swinging patio doors, skylights and roof-mounted (i.e., sloping) windows, and window systems that can be added to a house to create bay or bow windows, miniature greenhouses, or full sun rooms.

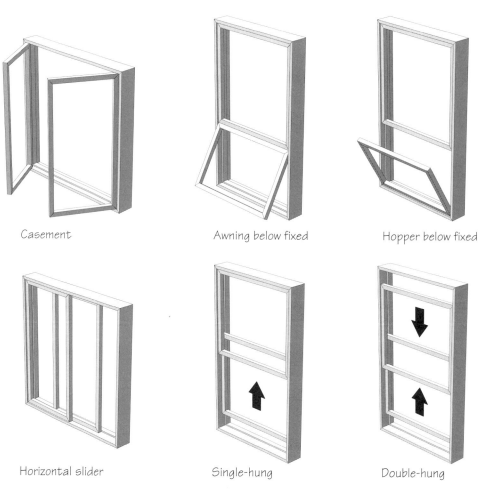

| Casement | Awning below fixed | Hopper below fixed |
| Horizontal slider | Single-hung | Double-hung |

Figure 4-2. Common window operating types.

## Projected or Hinged Windows

Hinged windows include casements, awnings, and hoppers—hinged at the side, top, and bottom, respectively. Some manufacturers also make pivoting and combination windows that allow for easier cleaning of the exterior surfaces. Hinged windows, especially casements, project outward, providing significantly better ventilation than sliders of equal size. Because the sash protrudes from the plane of the wall, it can be controlled to catch passing breezes, but screens must be placed on the interior side. Virtually the entire casement window area can be opened, while sliders are limited to less than half of the window area.

## Sliding Windows

Sliders are the most common type of windows and include horizontal sliders and single-hung and double-hung windows. Ventilation area can vary from a small crack to an opening of one-half the total glass area. Screens can be placed on the exterior or interior of the window unit.

In double-hung or double-sliding units, both sashes can slide. In double-sliding units, the same net amount of glass area can be opened for ventilation as in single sliders, but it can be split between the top and bottom or two ends of the window for better control of the air flow.

## Sliding Glass Doors

Sliding glass doors (patio doors) are essentially big sliding windows. As extremely large expanses of glass, patio doors exaggerate all of the issues related to comfort and energy performance. Since the proportion of glass to frame is very high for a glass sliding door, the selection of high-performance glass can have significant benefits.

## French Doors and Folding Patio Doors

French doors and folding glazed doors are growing in popularity. A basic double French door consists of two hinged doors with no center mullion, resulting in a 5- to 6-foot-wide (1.5 to 1.8 m) opening. Folding doors are typically made of pairs of hinged doors, so that a double folding door with two pairs of doors can create an opening of 12 feet (3.7 m) or more.

Figure 4-3. Casement windows can create large expanses of glazing and can be opened fully to provide ventilation. (Photo: Marvin Windows & Doors.)

Figure 4-4. Double-hung windows can be opened at the top and bottom to ventilate a room. (Photo: Dayton Technologies.)

89

Figure 4-5. Sliding glass doors can provide almost uninterrrupted views. (Photo: Marvin Windows & Doors.)

Figure 4-6. Operable roof windows provide light as well as ventilation in high spaces. (Photo: Velux-America Inc.)

## Skylights and Roof Windows

The vast majority of skylights are permanently fixed in place, mounted on a curb above a flat or sloped roof. However, hatch-style skylights that can be opened with an extended crank, push latch, or remote control motor are becoming more common. Typical skylights have a domed profile and are often made of one or two layers of tinted or diffusing plastic.

A roof window is a hybrid between a skylight and a standard window. They have become increasingly popular as homeowners and designers seek to better utilize space in smaller houses by creating habitable rooms under sloping roofs. They are glazed with glass rather than plastic and are available with most of the glazing and solar control options of standard windows. Both fixed and operable versions are available, and the operable roof windows can be opened manually or by a motorized system. In addition, some manufacturers offer special venting mechanisms that allow some ventilation air flow without actually opening the window. Operable skylights or vents allow hot air that rises to the ceiling level to be effectively exhausted from the space.

## Greenhouse (Garden) Windows

Greenhouse windows, also known as garden windows, are typically prefabricated frame and glass kits that can be inserted into a new or existing window opening. They may include shelves for plants or simply be used as a means of creating a greater sense of spaciousness. Greenhouse windows generally have higher heat loss and heat gain than a regular window of the same size because they contain more glazing than a conventional window that fills the same wall opening.

## Sun Rooms and Solariums

Sun rooms and solariums are glazed spaces attached to a house that are used for sitting or eating areas as well as growing plants. Sometimes they are furnished in the form of prefabricated kits and have the appearance of a greenhouse. They may also be built with the same construction and window types found in the rest of the house. Sun rooms and solariums may be fully heated and air-conditioned living spaces, or they may be used only seasonally where they are either semi-conditioned or unconditioned. Because they contain such a large amount of glazing area, it is essential to select efficient solutions, especially if the space is fully air-conditioned.

## Jalousie Windows

Jalousie windows are louvered windows typically installed in porches or other areas open to the outdoors for rain and wind protection. Because they do not seal completely, they are not appropriate for spaces that are heated or mechanically cooled.

## Combining Windows for Special Effects

In the simplest form, individual windows can be placed top to bottom and side by side to create the feeling of one larger expanse of window area. This results in a much greater feeling of connection to the outdoors, with more extensive views and daylight than an individual unit provides. Window manufacturers also provide special window configurations such as bay or bow windows. Different window operator types may be combined as needed—for example, fixed windows to

Figure 4-7. Efficient glazing materials are important in sun rooms and solariums. (Photo: Pella Windows.)

Figure 4-8. The traditional bay window is one way to combine smaller windows to enhance light and view. (Photo: Andersen Windows.)

enhance light and view with operable windows to provide ventilation and emergency egress. Large window walls may require special features to ensure structural integrity. These groupings of windows provide a feeling of openness and light as well as helping to define the architectural character of the exterior and interior of the house. With the almost unlimited array of window types and sizes, designers can create window combinations that are innovative and create the architectural character and spatial relationships they desire.

## Performance Implications of Basic Window Types

There are subtle performance differences between a fixed and operable window that fills an identical rough opening. The fixed unit will typically have a smaller fraction of frame and proportionately more glass than the similar operable unit. Thus, fixed windows with high-performance glass will have a better, lower U-factor, but a higher SHGC due to a smaller frame area and larger glass area. Fixed windows have very low infiltration rates, but then they also do not provide natural ventilation and do not satisfy building code requirements for fire egress.

For operating windows, the type of operation has little direct effect on the U-factor or SHGC of the unit, but it can have a significant effect on the air infiltration and ventilation characteristics. Window operation can be broken into two basic types: sliding windows and hinged windows. The comments below are a general characterization of American window stock; however, they may not apply to a specific window produced by a given manufacturer.

### Hinged Windows

Hinged windows such as casements, awnings, and hoppers generally have lower air leakage rates than sliding windows from the same manufacturer because the sash closes by pressing against the frame, permitting the use of more effective compression-type weatherstripping. In most types, the sash swings closed from the outside, so that additional external wind pressure tends to push the sash more tightly shut. Hinged windows require a strong frame to encase and support the projecting sash. Also, because projecting-type sashes must be strong enough to swing out and still resist wind forces, the stiffer window units do not flex as readily in the wind. In addition, hinged windows have locking mechanisms that force the sash against the weatherstripping to maximize compression. These design details tend to reduce air infiltration of hinged windows in comparison to sliders.

## Sliding Windows

Sliding windows, whether single-hung, double-hung, or horizontal sliders, generally have higher air leakage rates than projecting or hinged windows. Sliding windows typically use a brush-type weatherstripping that allows the sash to slide past. This type is generally less effective than the compression gaskets found in projecting windows. The weatherstrip effectiveness also tends to be reduced over time due to wear and tear from repeated movement of the sliding sash. The frames and sashes of sliding units can be made with lighter, less rigid frame sections since they only need to support their own weight. This lightness may permit the sliding frames to flex and can allow more air leakage under windy conditions. Manufacturers can choose to engineer greater stiffness in their products by design and material selection.

Slider window performance can also be improved with latching mechanisms that compress the sash to the fixed frame and by the addition of compression weatherstripping at the head and sill of double-hung windows or the end jamb of horizontal sliders.

---

**Effect of Window Type on Air Leakage**

Hinged windows (casements, awnings, and hoppers) generally have lower air leakage rates than sliding windows (horizontal sliding and double-hung windows). A compressive latch increases the effectiveness of the weather-stripping in preventing air leakage through hinged windows.

**Effect of Window Type on Ventilation**

Hinged units can provide natural ventilation through the entire window area. Sliding units can provide ventilation only through half the area or less. Double sliding units provide more control of ventilation than single sliders.

---

*Sliding Glass and French Doors*

As previously noted, sliding doors are essentially big sliding windows. However, they are more complicated because of their size and weight and because the sill is also a door threshold, which must keep water out while allowing easy passage of people and objects. The threshold is typically the most difficult part of the frame to weatherstrip effectively.

French doors benefit from being much more like traditional doors than sliding doors. French doors can use weatherstripping and operating hardware designed for similar nonglazed doors. However, when there are large openings with multiple hinged doors, it is more difficult to positively seal the joints between door leaves and to create the stiffness that will resist infiltration.

## Weatherstripping

Weatherstripping is an essential component of the operable part of a window. It must be able to flex each time the window is opened, and return to its original shape each time the window is closed. The quality of weatherstripping on a window is one of the main factors that distinguishes the quality of the window. Cheaper windows tend to save on cost by using poorer-quality, less expensive weatherstripping. Ideally, weatherstripping must survive thousands of operational cycles and years of exposure to sun, temperature change, and water without serious degradation of performance.

There are two basic strategies for weatherstrip design: brush or wiper types and compression types. Brush weatherstripping, made of felt, nylon, or polyester brushes, or synthetic rubber wipes similar to the wiper blades on a car windshield, sweeps against the window sash as it moves. Compression weatherstripping squeezes and expands with window operation. Materials used for compressible weatherstripping include felts, springy metal or plastic strips shaped into V-flaps, and synthetic rubber gaskets.

Each material has some disadvantages to be considered. Organic felts age fairly quickly, and all felts absorb moisture, reducing their effectiveness. Brush or wiper-type weatherstripping eventually gets matted down like a carpet that has had too much traffic. Metal strips are easily dented or bent and eventually lose their ability to recover their shape. Early synthetic plastics and rubber aged quickly, becoming brittle or sticky. Recently, progress has been made in developing synthetic plastic weatherstripping that is more durable and can be expected to last longer.

Window air leakage properties are measured under controlled environmental conditions as described in Chapter 2. These properties are typically determined for new windows under conditions that are not as severe as those the window will experience in a building. The window industry has developed additional test procedures that account for a wider range of environmental conditions, but these tests are more time consuming and costly, and are not widely utilized. Actual performance may thus vary from the rated properties for many reasons. The air leakage may be different at different environmental conditions—e.g., very cold temperature may distort a window shape or change the properties of the weatherstrip. Air leakage will change over time as materials age, or as a wall settles and distorts a window frame. Air leakage properties of a particular window can be tested in an existing house using an apparatus that pressurizes the window and measures the resulting air flow. A simpler, less quantitative technique involves the use of "smoke pencils," devices that generate a small visible stream of nontoxic smoke that can be introduced at the edge of a window to make visible the resultant air flow from infiltration. This can be used on a windy day, or the entire house can be pressurized to induce air flow.

From the perspective of comfort and energy use, the ability of the window to limit air leakage is an important performance capability. Window designers must select the right weatherstrip designs and materials to minimize air leakage, without adverse effects such as increasing the force required to open or close a sliding sash.

## Emergency Egress and Security

Windows have long been used as alternate escape routes during emergencies, especially during fires. Recognizing this, most codes regulate the size of free openings in windows, which must allow a person to escape from a bedroom or other "living area" or permit a fire fighter to enter.

While emergency egress from the inside is desired, most people want to restrict the entry from the outside. Latches should be quick and easy to operate from the inside, but

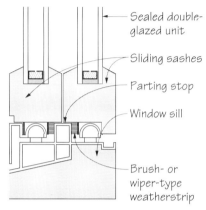

Figure 4-9. Brush- or wiper-type weatherstripping is suitable for sliding windows.

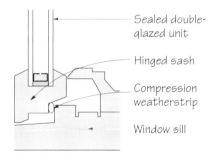

Figure 4-10. Compression-type weatherstripping is typically used for hinged windows.

difficult to see or reach from the outside. Screen or security bars must not restrict egress. Special security glazings, such as tempered and laminated glass or some plastic glazings, cannot be as easily broken as conventional glass and thus delays illegal entry. Operating windows can be wired into security systems to signal when the unit is opened. Windows can also incorporate sensors that report vibration or breakage to a monitoring unit.

## FRAME MATERIALS

At the beginning of the century, most residential windows were built on site by local carpenters. Millwork suppliers then started to manufacture window sash and even entire window units. Prefabricated steel-framed "factory" windows became popular for a while in the 1920s. Prefabricated wood-framed windows finally became big business with the tract home building that started during World War II and mushroomed in the 1950s. After the war, aluminum manufacturers turned their plants to domestic production and quickly found a large

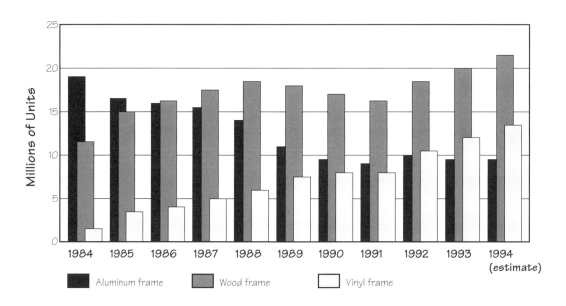

Figure 4-11. Total residential window sales by frame material (includes both new construction and remodeling). (Source: AAMA 1995.)

market with prefabricated, aluminum-framed windows. Until the mid-1980s, well over one-half of all residential windows in the United States were aluminum-framed. As illustrated in Figure 4-11, aluminum frames have been losing market share, initially to wood-framed windows and more recently to a relative newcomer, vinyl frames. During the 1990s new frames have been introduced made from fiberglass and engineered thermoplastics and new composite materials.

The material used to manufacture the frame governs the physical characteristics of the window, such as frame thickness, weight, and durability, but it also has a major impact on the thermal characteristics of the window. Increasingly, manufacturers are producing hybrid or composite sash and frames, in which multiple materials are selected and combined to best meet the overall required performance parameters. Thus, a simple inspection of the inner or outer surface of the frame is no longer an accurate indicator of the total material or its performance.

Figure 4-12 indicates the U-factors of various standard frame types. Since the actual U-factor depends on specific dimensions and design details, these values are only for illustrative purposes. While it is useful to understand the role that frame type plays in window thermal performance, the frame U-factor is not normally reported by manufacturers. The window U-factor, as given on an NFRC certified rating or label, incorporates the thermal properties of the frame. Since the sash and frame represent from 10 to 30 percent of the total area of the window unit, the frame properties will definitely influence the total window performance.

The remainder of this section describes wood, aluminum, and vinyl frames, and introduces some new frame materials that are becoming commercially available.

## Wood Frames

The traditional window frame material is wood, because of its availability and ease of milling into the complex shapes required to make windows. Today, wood units tend to be thought of as "high end" windows because competing products are often less expensive. Wood is not intrinsically the most durable window frame material, because of its susceptibility to rot, but well-built and well-maintained wood windows can have a very long life.

| Frame Material | U-factor |
|---|---|
| Aluminum (no thermal break) | 1.9–2.2 |
| Aluminum (with thermal break) | 1.0 |
| Aluminum-clad wood/ reinforced vinyl | 0.4–0.6 |
| Wood and vinyl | 0.3–0.5 |
| Insulated vinyl/ Insulated fiberglass | 0.2–0.3 |

Figure 4-12. Typical U-factors for frame materials.

**Characteristics of Wood Frames**

- Good thermal performance.

- Easy to mill into complex shapes suitable for windows.

- Exterior paint and maintenance required unless clad with vinyl or aluminum.

- Attractive interior appearance.

**Application of Wood Frames**

- All residential buildings.

Wood is easy to repair and maintain with simple tools and materials well understood by the average householder. A coat of paint protects the surface and allows an easy change in color schemes. Frames are generally made of kiln-dried lumber to reduce shrinkage, which results in better operation and weathertightness over their lifetime. Water-repellent and/or chemical treatments can be applied in the factory to reduce swelling and warping, improve paint retention, and increase wood's resistance to decay and insect attack.

Wood windows are manufactured in all configurations, from sliders to swinging windows. Wood is favored in many residential applications because of its appearance and traditional place in house design.

A variation of the wooden window is to clad the exterior face of the frame with either vinyl or aluminum, creating a permanent weather-resistant surface. Clad frames thus have lower maintenance requirements, while retaining the attractive wood finish on the interior.

Newer vinyl and enameled metal claddings offer much longer protection to wood frames. However, they are generally available in a limited number of colors, and refinishing these surfaces may be a difficult task. Dark-colored finishes absorb more of the sun's energy, so they tend to be more susceptible to aging from heat and ultraviolet damage. Many vinyl and plastic finishes and frame materials require special solvents and bonding agents for repair or refinishing. Given the specialized nature of repairing or refinishing many of these low-maintenance frame materials, it is best to contact the manufacturer to determine if repairs are feasible or replacement is the preferred option.

From a thermal point of view, wood-framed windows perform well. The thicker the wood frame, the more insulation it provides. Wood-framed windows typically exhibit some of the lowest U-factors in the residential marketplace.

However, metal cladding, metal hardware, or the metal reinforcing often used at corner joints can lower the thermal performance of wood frames. If the metal extends through the window from the cold side to the warm side of the frame, it creates a thermal short circuit, conducting heat more quickly through that section of the frame.

## Aluminum Frames

After World War II, aluminum became one of the most common residential window frame materials. Light, strong, durable, and easily extruded into the complex shapes required for window parts, it can be fabricated to extremely close tolerances, to create special forms for the insertion of glazing, weatherstripping, and thermal breaks. Aluminum frames are available in anodized and factory-baked enamel finishes that are extremely durable and low-maintenance. Aluminum is widely used for storm window units because of its light weight and corrosion resistance, and for sliding glass doors because of its strength.

The biggest disadvantage of aluminum as a window frame material is its high thermal conductance. It readily conducts heat, greatly raising the overall U-factor of a window unit. Because of its high thermal conductance, the thermal resistance of an aluminum frame is determined more by the amount of surface area of the frame than by the thickness or the projected area, as with other frame materials. Thus, an aluminum frame profile with a simple compact shape will perform much better than a profile with many fins and undulations.

In cold climates, a simple aluminum frame can easily become cold enough to condense moisture or frost on the inside surfaces of window frames. Even more than the problem of heat loss, the condensation problem spurred development of a more insulating aluminum frame.

The most common solution to the heat conduction problem of aluminum frames is to provide a "thermal break" by splitting the frame components into interior and exterior pieces and use a less conductive material to join them (Figure 4-13). There are many designs available for thermally broken aluminum frames. The most prevalent technique used in residential windows is called "pouring and debridging." The window frame part is first extruded as a single piece with a hollow trough in the middle. This is filled with a plastic that hardens into a strong intermediate piece. The connecting piece of aluminum is then milled out, leaving only the plastic to join the two halves of aluminum. Functionally, the resulting piece is cut, mitered, and assembled like a simple aluminum extrusion. Thermally, the plastic slows the heat flow between the inside and outside. There are other manufacturing techniques for producing a thermal break, but the thermal results are similar.

Current technology with standard thermal breaks has improved aluminum frame U-factors from roughly 2.0 to about half of that. This is difficult to achieve throughout the entire product, though. While easier to insert in the bulkier main frame, it is problematic to include thermal breaks in the thinner sash components that hold the glazing. In addition, as glazing systems with higher insulating values are developed, a frame with a U-factor of 1.0 or above may not be consistent with the energy efficiency of the other window components. However, innovative new thermal break designs combined with changes in frame design have been used in Europe to

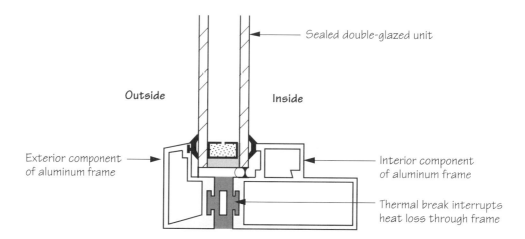

Figure 4-13. A thermal break in an aluminum frame.

> **Characteristics of Aluminum Frames**
>
> - High thermal conductance resulting in poor U-factors. Thermal breaks are necessary in most climates.
>
> - High strength-to-weight ratio.
>
> - Lightweight compared to wood.
>
> - Durable and low maintenance.
>
> **Applications of Aluminum Frames**
>
> - Used in climates where high thermal resistance is not a priority. Thermally broken frames are recommended to improve performance.
>
> - Used where lower maintenance and lower cost are requirements.

achieve U-factors lower than 0.5, but at a higher cost than current thermally broken frames. In hot climates, where solar gain is often more important than conductive heat transfer, improving the insulating value of the frame can be much less important than using a higher-performance glazing system.

The market share of aluminum windows has been steadily declining since the early 1980s, from a high of 58 percent in 1984 down to 22.5 percent in 1993. This reduction in market share has been seen for both new construction and renovation projects. Many of the companies that fabricated aluminum windows have switched to fabricating vinyl windows.

## Vinyl Frames

Plastics are relative newcomers as window frame materials in North America. Vinyl, also known as polyvinyl chloride (PVC), is a very versatile plastic with good insulating value, high impact resistance, and excellent resistance to abrasion, corrosion, air pollutants, and termites. Because the color goes all the way through, there is no finish coat that can be damaged or deteriorate over time. Recent advances have improved dimensional stability and resistance to degradation from sunlight and temperature extremes.

Developed and marketed for more than thirty years in Europe, vinyl frames first became available in the U.S. in the 1960s. In 1983, vinyl windows constituted about 3 percent of all residential window sales in the United States, mostly for remodeling and replacement. Since then, the use of vinyl windows has risen steadily in new construction and more dramatically in the remodeling and replacement market. By 1994, total vinyl window sales were conservatively estimated at 13.6 million units or about 30 percent of the total industry (AAMA 1995).

Similar to aluminum windows, vinyl windows are fabricated by cutting standard lineal extrusions to size and assembling the pieces into complete sash and frame elements. A small number of companies produce the standard lineal extrusions, and a much larger number of companies assemble window units by combining the sash and frame with glazing, weatherstripping, and hardware. This process lends itself to readily fabricating the custom sizes needed for the replacement window market.

Vinyl window frames require very little maintenance. They do not require painting and have good moisture resistance. To provide the required structural performance, vinyl sections are larger than aluminum window sections, closer to the dimensions of wood frame sections. Larger vinyl units may incorporate metal stiffeners.

While vinyl has a higher coefficient of expansion than either wood or aluminum, vinyl window frame profiles are designed and assembled to eliminate excessive movement caused by thermal cycles. Vinyl frames with heat-welded joints are stiffer than mechanically joined vinyl frames, so they provide excellent resistance to temperature and handling stresses both at the job site and during shipping. In some cases, a vinyl window assembly consists of both welded and mechanically fastened components. Interior webs also strengthen the frame, while improving its thermal performance. AAMA (American Architectural Manufacturers Association) has developed standards for vinyl extrusions to ensure impact resistance, dimensional stability, and color retention.

The thermal performances of vinyl frames range from comparable to wood to significantly better than wood. Large hollow chambers within the frame can allow unwanted heat transfer through convection currents. Creating smaller cells within the frame reduces this convection exchange, as does adding an insulating material (see Figures 3-44 and 3-45 in Chapter 3 for a comparison of vinyl frame thermal perfor-

---

**Characteristics of Vinyl Frames**

- Thermal performance depends on design features. Generally, energy performance is improved by separating cavities to prevent convection around the frame and by filling cavities with insulating material.

- Can be fabricated to custom sizes.

- Low maintenance.

**Applications of Vinyl Frames**

- All new residential buildings.

- Retrofitting.

---

mance). Most manufacturers are conducting research and development to improve the insulating value of their vinyl window assemblies.

New energy codes in some western states have spurred the sales of vinyl frames in those areas. These codes specifically require windows that have low overall U-factors, giving wood and vinyl frame windows a competitive advantage over metal frames. Given the recent rapid market penetration of plastic frames, much more development in the area of insulating frames is expected.

## Fiberglass and Engineered Thermoplastics

Vinyl windows have captured a large share of the new and retrofit window market in the last fifteen years in part because of the ability to produce a wide range of lineal extrusions that can be assembled into windows. Two other polymer-based technologies are beginning to challenge this market, although to date they have captured only a small market share. Windows can be made of glass-fiber-reinforced polyester, or fiberglass, which is pultruded into lineal forms and then assembled into windows. These frames are dimensionally stable and have good insulating value by incorporating air cavities (similar to vinyl). Because the material is stronger than vinyl, it can have smaller cross-sectional shapes and thus less area. Fiberglass windows are typically more expensive than vinyl windows. Another approach is to use extruded engineered thermoplastics, another family of plastics used extensively in automobiles and appliances. Like fiberglass, they have some structural and other advantages over vinyl but are also more expensive and have not yet captured a large market share.

## Wood Composites

Most people are familiar with composite wood products, such as particle board and laminated strand lumber, in which wood particles and resins are compressed to form a strong composite material. The wood window industry has now taken this a step further by creating a new generation of wood/polymer composites that are extruded into a series of lineal shapes for window frame and sash members. These composites are very stable, and have the same or better structural and thermal properties as conventional wood, with better moisture resistance and more decay resistance. They can be textured and stained or painted much like wood. They were initially used in critical elements, such as window sills and thresholds in sliding patio doors, but are now being used for entire window units. This approach has the added environmental advantage of reusing a volume of sawdust and wood scrap that would otherwise be discarded.

## Hybrid and Composite Frames

Manufacturers are increasingly turning to hybrid frame designs that use two or more of the frame materials described above to produce a complete window system. The wood industry has long built vinyl- and aluminum-clad windows to reduce exterior maintenance needs. Vinyl manufacturers and others offer interior wood veneers to produce the finish and appearance that many homeowners desire. Split-sash designs may have an interior wood element bonded to an exterior fiberglass element. We are likely to see an ever-increasing selection of such composite designs as manufacturers continue to try to provide better-performing products at lower cost. It may be important for a homeowner to learn about these materials from the perspective of maintenance requirements and options for interior finishes. However, it becomes increasingly difficult to estimate the thermal properties of such a frame from simple inspection. The best source of information is an NFRC label that provides the thermal properties of the overall window.

# INSTALLATION

The energy efficiency of a window unit, no matter how advanced the glazing materials or frame, also depends upon the quality of its installation and of any window treatments added to its exterior or interior. There are important differences in the details of how a window is installed, depending on the type of construction (wood versus masonry) or exterior cladding material (i.e., wood siding, stucco, brick veneer). In addition, each frame material type and each individual manufacturer has its own recommended installation practices. It is important to refer to the appropriate product information for your specific wall type and not to rely solely on general guidelines. Given the importance of proper installation, however, major window associations have also created standard guidelines for installation. Window installation guidelines are available from several sources listed in Appendix D. Some general conditions and principles are described below.

## Wall Framing Details

Windows are always placed into openings in walls that are sized slightly larger than the window unit itself. This additional space around the window, typically 1/4 to 1/2 inch (6 to 12 mm) on each side, is essential in providing construction tolerances and allowing for any movement of the building over time. However, it also creates a gap between the careful detailing of the window unit and the insulation of the wall.

As window insulating values improve, attention should focus on how to maintain a continuous insulating envelope between window and wall. As much care should be given to preventing infiltration and conduction heat losses at the joint between window unit and wall as is given to insulating the wall or selecting the window unit. The two most important ways to cut heat loss at the perimeter of a window are to place insulation into all voids and to use sealants to close off even the most minute air paths. Special attention must be given to the mounting of thermally broken aluminum frame windows, so that the integrity of the thermal break is not compromised.

Many window units are attached to the building with a fin mounting system; a casing placed over the fin provides a visual finish from the exterior. At this joint between the inside and the outside of the building, there is minimum thickness of material and a considerable air space that should be filled

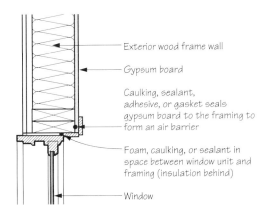

Figure 4-14. Jamb detail using gypsum board air barrier system.

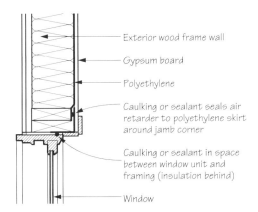

Figure 4-15. Jamb detail using polyethylene air barrier system.

with an insulating material. Fiberglass insulation can be manually stuffed into the void, or a foaming insulation can be shot in to provide a tight fit. Too much foam, however, can exert undue pressure, distorting the window frame and causing even greater air infiltration through the window. Some foam insulations that also act as sealant materials are available.

Innovative wall framing approaches can provide additional insulation levels around windows. Headers across the top of a window opening can be fabricated of two members spaced an inch (2.5 centimeters) or more apart, with the space between filled with insulation. Similarly, the jamb and sill framing can be detailed to allow additional insulation close to the opening. Stiffer methods of construction using 2 x 6 framing and/or structural sheathing allow for fewer framing members and, thus, more insulation right at the opening. In addition, if a rigid insulation layer is used, such as foam insulation around the interior or exterior of a masonry wall, the insulation should be wrapped into the opening to prevent any uninsulated edges of the wall from creating a "thermal bridge."

## Air and Vapor Retarders

Any infiltration or vapor retarder used in residential construction should maintain its integrity at the window opening, as illustrated in Figures 4-14 and 4-15. An interior plastic vapor barrier can be laid continuously across a window opening and cut on the diagonals to form an X cut at the window. The triangular edges are wrapped into the framed opening before the window is inserted. Alternatively, an exterior infiltration barrier, such as tar paper, should be folded into the window opening to make friction contact with the window unit. Caulking then creates a continuous seal from the window unit to the infiltration and/or vapor barrier.

Caulking to form a continuous vapor barrier is essential for extreme climates, either very cold, or very hot and humid. Even when condensation has been visibly eliminated from window glazing and frame surfaces by selecting thermally improved products, it can still cause problems. Water vapor will migrate unnoticed from high-pressure areas to low-pressure areas. If a hole in the vapor barrier envelope of a house happens to allow the water vapor to contact a cold surface, it will condense on the cold surface and collect there. If this surface is in the interior of a wall or between a window frame and a wall, the moisture can cause rot or rust damage, or render insulating materials ineffective. Such a condition can go undetected for years.

## Skylight Framing

Skylights and roof windows present a special case for insulating around windows because they are typically set into the thickest, most heavily insulated framing in the house, and they must also meet much more stringent conditions for shedding water.

In order to create a positive water flow around them, skylights are commonly mounted on "curbs" set above the roof plane (Figure 4-16). These curbs, rising 6 to 12 inches (15 to 30 centimeters) above the roof, create additional heat loss surfaces, right where the warmest air of the house tends to collect. Ideally, they should be insulated to the same level as the roof. In practice, it is often difficult to achieve insulation levels much above R-11. Some manufacturers provide curbs prefabricated out of a rigid insulating foam, which can be further insulated at the site.

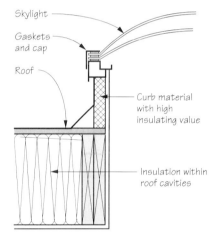

Figure 4-16. Skylight detail with insulated curb material.

Roof windows, mounted in a sloping roof, often include a metal flashing system. If this metal flashing is in contact with a metal window frame, it can create additional surfaces for conducting heat. Thus, as with thermally broken aluminum windows, care should be taken to ensure a thermal separation between the cold outer metal surfaces and metal parts of the window frame that are exposed to the warm interior air.

## Installing Replacement Windows and Sashes

Replacement windows can be considered in two categories: remodeling or renovation windows, where the original window is removed and completely replaced, and replacement sash, where the original frame remains in place, and only new glass and operating sash is installed.

Remodeling and renovation windows should follow all of the considerations discussed above for new windows. Insulation and infiltration barrier continuity should be maintained all around the window. Foam-in-place sealants can be used to fill irregular voids created between old and the new components; however, as with new installations, too much foam can distort the frame and have detrimental effects on performance. Remodeling is often an opportunity to check the interior condition of walls and increase insulation in the opened areas.

Replacement windows often make use of a hollow *panning* frame to fit around existing odd-shaped exterior frame members. This typically aluminum or vinyl frame provides weather protection for the existing frame and a visually continuous element. However, it can also conceal hidden problems. Careful caulking around the panning frame is essential. The voids behind it are also good candidates for nonmoisture-retaining insulation, such as foam-in-place.

Replacement sashes involve less expense and disruption for a household. They are custom-sized and detailed to fit into existing window frames. Only the glazing and operable sash are new. This is a good approach for upgrading window energy performance when the original window frames are in good shape. At the same time the sash is replaced, new weatherstripping that is most appropriate to the window type and frame details should be installed. Some of the benefits of energy-efficient glazing can be compromised if the new sash is not properly weatherstripped.

**Window Installation Guidelines**

- Carefully insulate all voids left between window and wall.

- Caulk or seal around the perimeter of windows to maintain the integrity of air and/or vapor barriers.

- For replacement windows, insulate all voids and caulk all joints between old wall components and new window assembly.

- Set windows to maintain the integrity of the thermal break in the window frame.

- Insulate skylight curbs at levels similar to wall insulation.

- Maintain a thermal break for the installation of metal frame skylights.

Some replacement window manufacturers offer a compromise between complete window replacement (where the cost of removing the entire old window frame is high), and a simple sash replacement. The old sash and other adjacent trim is removed, leaving the original frame in place. The new window is inserted into this framed opening, following prescribed installation procedures. Then, appropriate trim is installed on the interior and exterior. This provides many of the benefits of a complete window replacement, but at a lower cost. However, the net effect is typically to reduce the total glazing area since a complete window assembly is basically fitted into the old frame.

In a case where the existing windows are architecturally or historically unique, special care should be taken with renovation or replacement. It is generally possible to incorporate some or all of the desirable energy efficiency improvements in such cases, although the options may be reduced if the renovated windows must meet specific architectural or historical requirements. Consult appropriate specialists or custom window manufacturers with experience in this field.

# CHAPTER 5

# Design Implications with Energy-Efficient Windows

It is clear that the advanced window technologies described throughout this book can have a major effect on comfort and on the annual energy performance of a house. However, there is a broader and possibly more significant impact of the recent revolution in window performance. Because the new glazing technologies provide highly effective insulating value and solar protection, there are important implications for how a house is designed.

There has been a long-established set of window design guidelines and assumptions intended to reduce heating and cooling energy use. These are based, in part, on the historical

Figure 5-1. High-performance windows make energy-efficient homes possible with greater freedom of design than in the past.

assumption that windows were the weak link in the building envelope. These assumptions frequently created limitations on design freedom or generated conflicts with other performance requirements, such as view. Traditional considerations include orientation, amount of glazing, and shading requirements for windows. The new technologies, however, raise questions about the validity of many of these restrictive assumptions. It can be argued that many assumptions developed over the last twenty years about energy-efficient home design require rethinking in light of the new window technologies.

Windows are one of the most multifaceted components of a home. They affect the aesthetics of the home, provide for views, ventilation, and daylight, and have a major impact on the comfort, energy consumption, maintenance requirements, and cost of the home. The designer must determine the amount and placement of glazing to provide natural ventilation in summer, solar heat gain in winter, or simply to allow egress from bedrooms as required. Decisions about window size and placement are also integral to the exterior appearance of the home and the interior aesthetic qualities of the spaces. Possible design objectives include creating a sense of spaciousness and providing natural light or particular views. These architectural decisions are interwoven with site design. Window placement to provide light, solar heating, or ventilation can be enhanced or diminished by landscape elements.

The entire range of window selection considerations is summarized in Chapter 6; this chapter addresses the relationship between window selection and building design. To develop a useful set of guidelines for window sizing, placement, and other building design concerns, it is necessary to consider the multiple design objectives related to windows. These are:

- Providing views.

- Providing daylight.

- Providing fresh air.

- Decreasing summer heat gain.

- Decreasing winter heat loss.

- Providing winter solar heat gain.

The first three of these—providing views, light, and fresh air—represent traditional window functions that do not change fundamentally as new technologies are introduced, although the ability to provide more daylight, for example, without a significant energy penalty is now available. However, the second three objectives—decreasing summer heat gain, decreasing winter heat loss, and providing winter heat gain—are influenced considerably by the ability to select high-performance windows. Guidelines for achieving each of these objectives are discussed in the remainder of this chapter.

## PROVIDING VIEWS

When people think of windows, often the immediate association is not with the window unit itself, but rather with the view provided by the window. Views that are highly valued, such as views of the ocean, lake, trees, or mountains, often involve subtle movement and changes in light throughout the day, which can be both mentally restful and stimulating. The view out of a window also contributes to our sense of orientation. Glimpses of familiar scenes or landmarks give a sense of place in the environment and within a building.

---

**Guidelines for Providing Views**

- Locate and size windows to take advantage of attractive and interesting exterior views and to maximize the connection between indoor and outdoor space.

- Develop attractive views with landscaping or courtyards when expansive views are not available.

- Size and place windows to frame views so that attractive elements are seen and undesirable features are blocked out.

- Place windows to provide surveillance of children's play areas or entry approach.

- Locate windows and use site design elements to maintain privacy.

- Use shades, curtains, or other operable devices to provide privacy when desired.

---

Windows provide a connection with the natural environment and a relief from typical interior spaces. An enormous amount of information can be gathered by a simple glance out a window: the time of day, weather conditions, and the coming and going of other people.

## Connecting Indoor and Outdoor Space

Using one larger window (or group of windows) instead of several smaller, scattered windows can have a powerful effect by visually connecting indoor and outdoor spaces. The picture window, which gained popularity with the availability of plate glass in the 1950s, made it possible to have a wide view without the distortion of many smaller windows. Sliding glass doors provide an even greater sense of openness with the possibility of floor-to-ceiling and wall-to-wall glazing. Creating a sun room that has glazing on two or three walls (and sometimes the roof as well) results in a space that feels as if it is at least partially outdoors.

As more homes are used for offices, visual relief is an important design consideration for a person sitting at a desk for hours. Exterior views do not have to be large and distant to have a very positive effect. A view of a small courtyard or even a simple garden wall with plantings can be quite attractive at close range.

Careful placement of windows allows for selection and modification of views. A larger window does not always provide a better view. A high sill can cut off the view of an unpleasant parking lot or road. A narrow window can cut the neighbors' house out of the view, leaving only their magnificent garden to be enjoyed. Of course, the position of the viewer on the interior must be considered when framing exterior views through a window.

## Using Windows for Security and Surveillance

In addition to providing attractive views, windows allow visual and verbal communication between the inside and outside of a home. From a window, you can see who is approaching your house, if the neighbors are home, or when the kids arrive home from school. Parents have always valued windows as a way to monitor their children playing outside. The window at the kitchen sink that allows supervision of the backyard while doing kitchen work has become a common pattern in the

Figure 5-2. Sun rooms provide a connection between indoors and outdoors.

Figure 5-3. View into courtyard.

home-building community. This visual supervision issue is extremely important. A wall of a house with no window in it becomes a "blind spot" in the yard, making supervision difficult.

### Maintaining Privacy

A window must provide for just the right level of privacy for the inhabitants. Because a clear glass window is a two-way street, privacy becomes an important issue. A window is one of the main filters between our private world and the public realm beyond. The careful location of a window can resolve many privacy issues: a small window or a window placed high in a wall will restrict the view to the interior, while allowing a view of the outside.

Tinted or reflective glass can provide some degree of privacy, but its effectiveness depends upon a bright exterior and a dim interior. At night, when the relative brightness is reversed, tinted glass no longer offers much privacy. Frosted glass, textured glass, glass block, and stained glass can prevent a full view from the outside while providing daylight, but they just as effectively prevent a view to the outside. Curtains and blinds allow a full range of individual control, and are the primary way most people create the right level of privacy for their homes to accommodate changing privacy needs.

New kinds of window glazing (e.g., the liquid crystal privacy glazing described in Chapter 3) allow the transparency of a window to be controlled electronically. The window is clear with the power on; switching the power off causes the optical material between two panes of glass to change its transparency. This obscured window, which is similar in appearance to frosted glass, still allows the passage of most light.

## PROVIDING DAYLIGHT

Letting light into a house is an important function of windows. Even though people have become more reliant on electric light in their houses, good home design can provide most, if not all, the needed daytime light. The qualities of natural light are important even though the amount of energy that can be saved with daylighting is less significant in houses compared to commercial buildings.

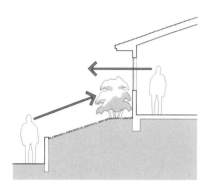

Figure 5-4. Landscape elements such as grade changes, retaining walls, and plantings can be used to ensure privacy while maintaining view out.

The introduction of natural light is also a powerful architectural tool in shaping and defining the interior spaces. However, providing daylight requires thoughtful window placement and interior design that addresses a number of concerns such as visual comfort, balanced light levels, color, and fading of furnishings.

## Providing Balanced Lighting

A balance of light is important both for visual comfort and to perform visual tasks. Too much contrast between dark and light, from very bright light to very dark shadows, can be uncomfortable for the eyes. Although the eyes can adjust to changes in light levels very quickly and can simultaneously see a wide range of light intensities, the human eye is more comfortable with a ratio of the brightest to the darkest level of no more than twenty to one. Light levels for reading in the home or office range from 100 to 2000 lux, and an overcast sky can provide outdoor levels of 5000 to 20,000 lux. A beam of bright sunlight will provide up to 100,000 lux on a surface. In order for direct sunlight to be useful visually, it should be diffused and reflected around the room. Most people have experienced the difficulty of reading in bright sunlight. When the sunlight is spread out over a larger area, it provides more comfortable light levels, at 100 to 2000 lux.

It is important to recognize that different uses of interior spaces have different ranges of acceptable lighting level. In a corridor, the amount of illumination can range above or below desired levels with little adverse impact. However, in a home office with a computer screen, visual comfort depends on careful control of brightness ratios. Artwork and artifacts, particularly those with paints, dyes, paper, or fabric that are light-sensitive or susceptible to fading, must also be protected from excessive light levels.

The balance of light in a space depends on the overall number and size of windows, their location, and the average reflectance of the interior surfaces of a room. A room with only one window will inevitably have bright areas close to the window and dark areas farther from the window. This gradation in light will be further exaggerated if the room has dark surfaces and furnishings. An improved balance of light can be created by providing light from at least two directions, such as windows located on different walls or a skylight balancing the

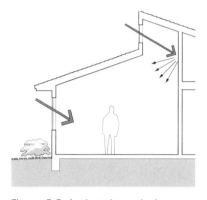

Figure 5-5. A clerestory window enhances architectural volumes with natural light. A light-colored wall acts as a reflector and a diffuser for daylight entering through the clerestory. This approach brings light into the space from two directions, which contributes to more balanced lighting levels.

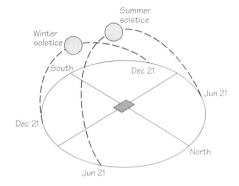

Figure 5-6. Solar path diagram illustrates the range of sun altitude and direction during the year.

light from a window. Shadows created from the first window source are balanced by light from the second direction, and the overall contrast is reduced.

As the sun moves through the sky during the day and in different seasons of the year, its angle of penetration into a room changes significantly. In the middle of the day during the summer in the northern hemisphere, for example, the sun comes in high overhead and strikes near the sill of a south-facing window. However, the low sun to the east in early morning and to the west during the afternoon will penetrate deep into a room, strike back walls, and shine directly into people's eyes. In the winter, the sun is lower in the sky

**Guidelines for Providing Natural Light**

- Arrange windows to provide daylight to all occupied rooms.

- Locate windows to define and enhance architectural volumes.

- Provide balanced lighting by introducing daylight from two directions in order to avoid glare and bright visual hot spots.

- Place windows so that direct sunlight, if admitted at all, reflects off interior walls and floors to provide more diffuse, even light.

- Use reflective ground surfaces or walls to increase day-light distribution into south- and north-facing windows.

- Avoid reflective ground surfaces that will increase glare from low sun entering east- and west-facing windows.

- Use translucent glazings on skylights to diffuse direct sun-light. Consider installing shutters or shades to block high midday summer sun while admitting daylight in morning, early evening, and on overcast days.

- Use landscape elements to block low direct sun into east- and west-facing windows.

- Use shades/curtains/overhangs to block sunlight from the sky.

- Use light from the north-facing windows to provide less variable, more diffuse illumination when desired.

throughout the day for all orientations, and tends to penetrate more deeply into rooms. This may provide useful heat but must be managed to control glare. In order to be visually useful, direct sunlight must first be reflected off of a floor or wall and then diffused around the room. For instance, east- or west-facing windows can be placed near a corner with a north wall of a room, which will catch and reflect the sunlight before it penetrates too deeply. In addition to designing the room itself properly for glare control, venetian blinds, translucent shades, and drapes can be used to diffuse entering light. In hot climates, it is usually best to avoid direct sun completely during the peak cooling season.

## Using Reflective Outdoor Surfaces to Increase Daylight

Direct light from the sun and sky is an important source of daylight, although in hot climates, it is usually best to avoid direct sunlight completely during the cooling season. In all climates there are situations in which the light reflecting off of outside surfaces can be very useful. A view of a light-colored sunlit wall can actually provide more light than a view of the north sky. And large reflective horizontal surfaces outside a window, such as snow, a lake, or a sandy beach, greatly increase the amount of light entering the window.

People often think of south windows as the sunny windows; however, the view from a south window might look out at the dark, shady "back" of a building, which is in stark contrast to the bright, sunlit foreground. On the other hand, a north window can look out onto a brightly sunlit wall or garden, providing both a great deal of reflected light and a cheerful view of sunlit flowers or surfaces.

## Avoiding Glare from the Sun

While sun penetration at south windows in winter can create some glare problems, east- and especially west-facing windows are usually the greatest offenders year round. People inevitably orient windows toward the most interesting view. But when that view includes a reflective surface such as snow, water, or sand, especially if the window faces the east or west horizon, the penetrating low sun problem is intensified. These windows create the most difficult situations for solar control.

Figure 5-7. Reflective ground.

While the view may be highly desired, the glare and excess heat of the sun and its reflections are not. Typically, these windows require active operation of shades or blinds by the homeowner, and, until recently, the only permanent solar control technology was tinted or reflective glazing or plastic films applied to the glass.

Direct sun at low angles can be blocked by trees, shrubs, or garden walls. Such landscape elements can be strategically placed to reduce glare as well as heat gain through east- and west-facing windows. The most important time to block low-angled sun is in the summer when the sun rises and sets farther to the north of direct east and west, so plantings should be located to account for this pattern.

## Diffusing Direct Sunlight

Overhangs and other solid external architectural elements can block direct sunlight completely for some time periods, while more open elements such as lattice structures diffuse the daylight before it enters the windows. This can assist in reducing glare from the direct sun or sky, and can illuminate spaces with a larger and more diffuse light source. Dark-colored woven fiberglass or perforated metal screens mounted on the exterior of windows can also reduce glare while still maintaining a high degree of visibility to the outside on sunny days. Light-colored screens diffuse the transmitted light but do not allow as much view to the outside.

Interior shades, drapes, or blinds can block or diffuse direct sunlight. The ideal drapery to reduce glare from bright windows, while still allowing a clear view out, is a loosely woven dark-colored drape. A loosely woven light-colored drape will diffuse the daylight from the window about the room and provide maximum privacy. However, light-colored drapes will appear very bright in direct sunlight and it may be difficult to see through them. Roll-up shades can be translucent, allowing some diffuse light to enter, or opaque, which block all light. Shades can also be made from woven screen materials in varying densities and colors, producing a range of light control and visibility. Reflective and tinted plastic roll-up shades reduce the sun's intensity and provide a clear view out, but do not diffuse or scatter the incoming light. When adjusted to the correct angle, horizontal venetian blinds or miniblinds can block direct sunlight but permit diffuse light to enter the room and often provide views out. Vertical blinds serve a similar function and are particularly useful for controlling low-angle sun at east- and west-facing windows.

Translucent glazing materials, such as frosted or patterned glass, can diffuse sunlight very evenly. Glass block and translucent fiberglass panels also provide light diffusion and visual privacy. The use of translucent glazings for skylights and clerestory windows, which are not in the visual field, is an excellent way to diffuse sunlight evenly throughout a space. When used on view windows, however, the materials may become too bright for visual comfort. A frosted window in the low afternoon sunlight will seem to glow with the intensity of a searchlight. Because of this effect, clear glass with additional interior shading devices generally allows for better control of the light from sunlit windows.

Skylights and roof windows can provide high levels of daylight, but view is not a concern as it is with conventional windows. The direct light from above can be diffused by using frosted glazing, by using a light well where light-colored vertical surfaces just beneath the skylight reflect and diffuse sunlight, or by using interior shades or blinds operated from below. A single skylight is far more effective at lighting a larger space on a sunny day if the sunlight is diffused either by the glazing or the light well surfaces. The size, shape, and color of light well surfaces influences their ability to diffuse and distribute light.

## Using Sky Light

Light from the sky, as distinguished from light from the sun, is cooler, gentler, and diffuse. While sunlight is normally considered to appear white, the early morning and late afternoon sun takes on a yellow or red hue, an effect that can be heightened by dust and pollutants in the atmosphere. Depending upon the position of the sun and type of clouds present, light from the sky will provide illumination levels of 5000 to 20,000 footcandles, or 5 to 20 percent of that provided by bright direct sunlight. Light from a clear blue sky has the additional advantage that there is more visible light, with a significantly smaller infrared component.

In situations where daylight is desired with minimal solar heat gain, north windows can provide the best quality of daylight of any orientation. Artists have long preferred the

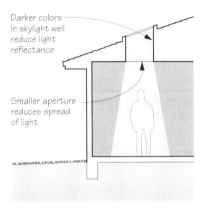

Figure 5-8. Roof window with smaller, dark-colored light well reduces amount of light and increases contrast.

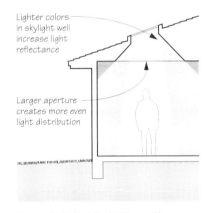

Figure 5-9. Roof window with larger, light-colored light well increases amount of light and provides more even distribution.

light from a high north window for their studios because the intensity, quality, and color of the light is most constant throughout the day. While the use of north-facing glass may be desirable in terms of daylighting and avoiding solar heat gain, it is not an effective strategy for providing useful solar gain in winter.

## PROVIDING FRESH AIR

Windows provide the primary means to control air flow in most homes. People open windows to provide fresh air, ventilate odors and smoke, dissipate heat and moisture, and create air movement on hot days. While exhaust fans and central air systems can mechanically ventilate a room, opening a room to the outdoors is perceived as more direct and natural. Architects have often found that people will insist upon having operable windows even though they may rarely open them. There seems to be a strong psychological need to know that a window *could* be opened, if necessary.

In order to ensure that all residences have access to the healthful aspects of natural ventilation, state or local building codes commonly regulate the minimum size of the ventilation opening in a window and the egress opening. Typically, codes require that about 5 percent of the floor area of a "living area," such as a bedroom or living room, be provided in ventilation area. These regulated areas should be carefully checked before sizing or replacing a window.

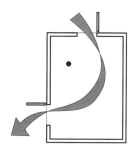

Figure 5-10. Casement windows can be used to deflect air into a room. Black dot indicates area of static air.

---

**Guidelines for Providing Fresh Air**

- Place operable windows in all rooms to give occupants opportunity for fresh air.

- Provide cross-ventilation by placing window openings on opposite walls in line with the prevailing winds.

- Use casement windows to direct and control ventilation.

- Use operable skylights or roof windows to enhance ventilation.

- Use landscape elements to direct breezes.

The potential value of natural ventilation as an energy efficiency strategy depends on climate and lifestyle. In a mild climate, there may be many hours during the year when outdoor air can be used to improve comfort and save energy for air conditioning. In climates with severe summers and/or winters or in dusty, noisy, or humid locations, the value of natural ventilation may be limited. To the extent that natural ventilation requires occupants to open and close windows, the lifestyle of the occupant may also be a factor. Finally, security concerns may limit the opportunities to provide natural ventilation.

An alternative approach that provides a steady amount of outdoor air is to incorporate a small ventilation element into the frame of the window. This passive approach has been used in the northwest United States to meet state code requirements for a minimum amount of outdoor air in new, tightly built houses that do not use mechanical ventilation. These "trickle ventilators" have been widely used in European houses but are relatively new in the U.S. The slots go through the window frame, normally on the top or bottom, with screens and flaps that keep out bugs and rain. They can be adjusted by occupants to control the amount of air flow. The peak amount of air exchange can be controlled in each room by properly sizing the ventilators in each window.

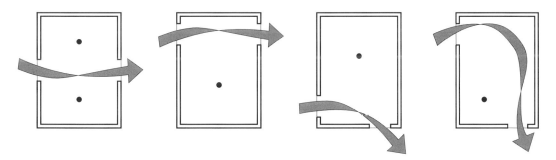

Figure 5-11. Location of windows determines the air flow path through a room. Black dots indicate areas of static air.

## Improving Ventilation

Good ventilation design provides for cross-ventilation of as many spaces in the house as practical. In normal wind conditions, the side of a building facing the wind will have a zone of positive pressure and the opposite side will have a zone of negative pressure. By providing adequate ventilation openings on these two sides of the house, a positive flow of air through the interior, from positive to negative pressure, is encouraged. Of course, the interior layout of the house must permit the air to flow through, and interior doors in the ventilation path must remain open.

If windows cannot be located on opposing walls, high- and low-pressure areas can be induced with the use of casements. Of all window types, casements provide the most control of ventilation direction and intensity. Because the sash can be opened into an air stream, breezes that would otherwise pass by can be directed into the room. Window types in which the sash remains flush with the wall ventilate well only with direct pressure differences across the window. In addition, as noted previously, virtually the entire window area of casement units can be opened, while sliders are limited to less than half of the area.

Even without external winds, double-hung windows can sometimes provide natural ventilation caused by stratified air flow within a room. Cooler fresh air enters at the bottom opening, while hotter air at the ceiling level is allowed to exit

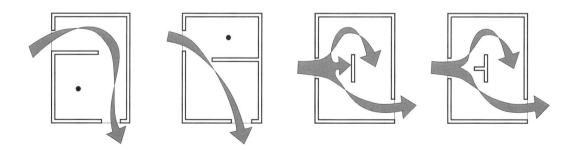

Figure 5-12. Partitions influence ventilation patterns. Black dots indicate areas of static air.

through the top opening. The taller the windows and the higher the ceiling, the more pronounced is this effect.

Operable skylights or roof windows can aid overall ventilation in a house significantly by creating a similar thermal chimney effect, letting hot air escape from the ceiling level where it accumulates and causing cooler air to be drawn in through lower windows (Figure 5-13).

Trees, shrubs, exterior walls, and earth berms can all divert wind patterns. These elements can be used to some extent to funnel or direct breezes through a house. Conversely, if these elements are placed without regard for prevailing wind patterns, they can act as obstructions to natural ventilation.

## DECREASING SUMMER HEAT GAIN

### Orienting Windows to Reduce Heat Gain

Until recently, windows themselves had little inherent capability to reduce solar heat gain, so the layout of energy-efficient houses evolved to protect windows from the most significant gains. In hot climates, the goal with this approach is to face most windows north, where there is little direct exposure, or to the south, where they can easily be designed with overhangs that will keep out most of the hot summer sun. Overhangs are much less effective against the lower angles of the east and west sun. Therefore, simply reducing the size and number of east and west windows can be the most direct strategy. West windows are subject to the full force of the strong afternoon sun, at a time of day when temperatures generally climb to their peak. East windows have the same problem in the morning hours, but air temperatures tend to be cooler at that time.

These traditional patterns are not necessarily valid, however, when better-performing windows with low solar heat gain coefficients are used. Figure 5-14 illustrates the impact of different window orientations on cooling energy use for a house in Phoenix, Arizona. As expected, facing windows in different directions has a significant impact when typical single- or double-glazed windows are used. When higher-performance windows with spectrally selective low-E coatings are used, however, the window orientation has a greatly diminished impact on energy use. In effect, with these more

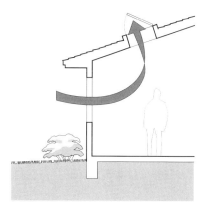

Figure 5-13. Operable skylights or roof windows can enhance natural ventilation.

Figure 5-14. Impact of window orientation on annual cooling
season energy use in Phoenix, Arizona.

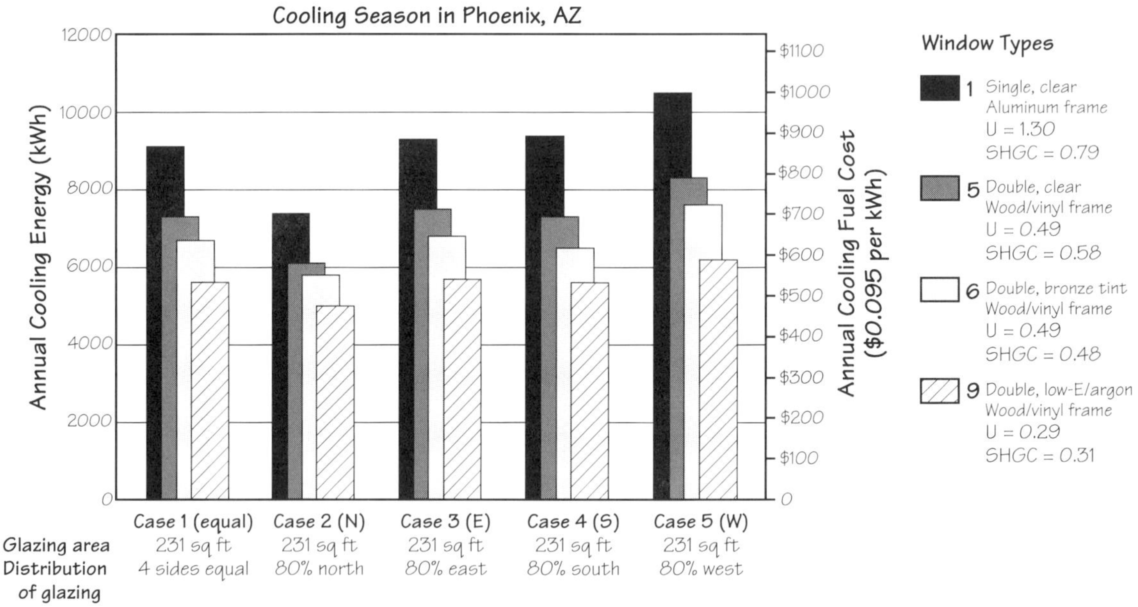

Note: The annual energy performance figures shown here are for a typical 1540 sq ft
house. U-factor and SHGC are for total window including frame. House and windows
are described in Appendix A. kWh=kilowatt hours.

advanced windows, nearly all of the glazing can face west or
south without a significant energy penalty. Moreover, these
computer simulations are done for a house with no overhangs
or external or internal shading devices. However, localized
glare and overheating can occur with this strategy, effects that
are not shown on these annual energy performance simulations.

## Reducing the Glazing Area to Reduce Gain

Another simple guideline to reduce heat gain is to reduce the
total glazing area. Of course, this can be effective with any type
of window, but it is particularly important when less efficient
windows are used. Figure 5-15 illustrates the impact of three

## Guidelines for Decreasing Heat Gain in Summer

Using clear single- or double-glazed windows:

- Design the layout so that windows and living areas do not face the hot western sun.

- Minimize window area on sunlit orientations to reduce solar heat gain.

- Use overhangs or other architectural and landscape elements to prevent sunlight from reaching windows.

- Use drapes, blinds, shades, or other interior treatments to reduce heat gain through windows.

Using high-performance windows (low SHGC):

- Orientation is less of a concern.

- Window area can be increased without a significant cooling energy penalty.

- Shading with external devices or landscaping is less of a concern.

- Internal shades are not as critical for solar control purposes.

    Note: Proper orientation and effective use of internal and external shading devices can still be useful in minimizing summer heat gain, but if they are not practical solutions for other reasons, low SHGC windows can largely compensate for their omission.

different amounts of glazing on cooling energy use for a house in Phoenix, Arizona. In all cases, the windows are equally distributed on the four orientations. With the single- and double-glazed windows, increasing or decreasing the glazing area from a typical glazing-to-floor-area ratio of 15 percent has an enormous impact on the cooling load. The cooling energy use for a house with higher-performance windows (low solar heat gain coefficient) still exhibits the same pattern, but the differences are not nearly as great in relative or absolute terms.

Because of the need for daylighting, views, and ventilation, reducing window area significantly is not a realistic or desirable strategy. This analysis indicates, however, that increasing glazing area for any reason will not have nearly as profound an impact when high-performance windows are used. When larger windows are desired, the combination of high-performance glazing plus appropriate exterior and interior shading devices can reduce the cooling impact to that of a smaller, unshaded window. Figure 5-15 also shows that a house with a 15 percent glazing-to-floor-area ratio using high-performance windows uses the same amount of cooling energy as a house with a 5 percent glazing-to-floor-area ratio using single-glazed windows. In effect, by using better windows, you can have three times as much glazing area with no energy penalty.

Figure 5-15. Impact of window area on annual cooling season energy use in Phoenix, Arizona.

Case 1: Glazing area is 5% of floor area (77 sq ft).

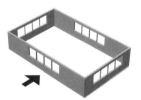

Case 2: Glazing area is 15% of floor area (231 sq ft).

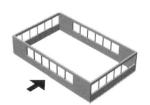

Case 3: Glazing area is 25% of floor area (385 sq ft).

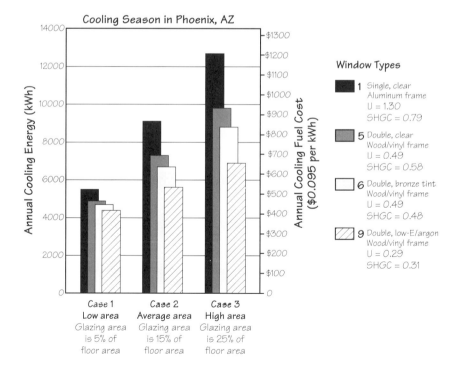

Note: The annual energy performance figures shown here are for a typical 1540 sq ft house. U-factor and SHGC are for total window including frame. House and windows are described in Appendix A. kWh=kilowatt hours.

## Using Overhangs and Exterior Shading Devices

Since ordinary windows have traditionally been the primary source of heat gain in summer, any effort to shade them has had great benefits in terms of comfort and energy use. The best place to shade a window is on the outside, before the sun strikes the window. Exterior shading devices have long been considered the most effective way to reduce solar heat gain into a home. The most common is the fixed overhang. For south-facing windows, overhangs can be sized to block out much of the summer sun but still permit lower-angled winter sun to enter. Figure 5-16 shows a simple procedure to estimate the size of a south-facing overhang designed to provide summer shading. Other exterior devices include grills, awnings, roll-down shutters, canopies, Bermuda shutters, and bamboo shades. The choice of shade type is often distinctly regional, based on local traditions.

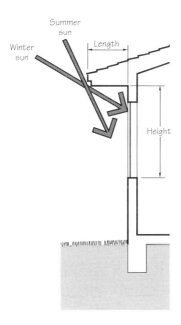

Figure 5-16. Overhangs can be used to shade summer sun while permitting beneficial winter sun to enter.

The following equation provides a quick method for determining the length of an overhang:

$$\text{Length} = \frac{\text{Height}}{\text{F}}$$

Using the table below, select an F-factor according to your latitude. The higher values will provide 100% shading at noon on June 21, the lower values provide 100% shading at noon from about mid-May until August 1.

| North Latitude | F-factor Provides full shade on 6/21 | F-factor Provides full shade from 5/10 to 8/1 |
|---|---|---|
| 28° | 11.1 | 5.6 |
| 32° | 6.3 | 4.0 |
| 36° | 4.5 | 3.0 |
| 40° | 3.4 | 2.5 |
| 44° | 2.7 | 2.0 |
| 48° | 2.2 | 1.7 |
| 52° | 1.8 | 1.5 |
| 56° | 1.5 | 1.3 |

(Source: Mazria, 1979.)

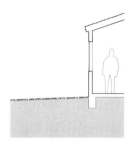

Case 1: Typical house with no shading.

Case 2: Typical house with overhangs—2 feet (0.6 m) deep, 1 foot (0.3 m) above window.

Case 3: Typical house with extensive tree shading.

Among the possible exterior shading devices are black or dark-colored screens mounted on the exterior of the window. Such a fabric or metal "solar screen" can lower a clear window's solar heat gain coefficient by 30 to 70 percent. The open weave of the screen allows much of the heat from this absorbed radiation to be convected away before it interacts with the window's glazing.

Reliance on external shading devices is not nearly as important, however, when windows with low SHGC are used. Figure 5-17 illustrates the impact of overhangs and a densely wooded lot on cooling energy use for a house in Phoenix, Arizona. As expected, the external shading devices reduce total cooling energy use by 20 to 25 percent when single glazing is used and by a significant amount for clear double

Figure 5-17. Impact of overhangs and shading on annual cooling season energy use in Phoenix, Arizona.

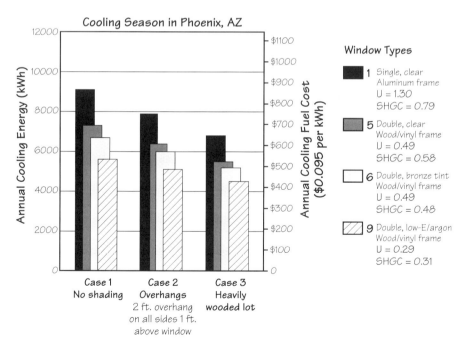

Note: The annual energy performance figures shown here are for a typical 1540 sq ft house. U-factor and SHGC are for total window including frame. House and windows are described in Appendix A. kWh=kilowatt hours.

glazing as well. Using spectrally selective low-E coatings (window 9 in Figure 5-17), however, results in a more modest benefit from the use of overhangs. This is because the glazing itself provides the necessary control of solar radiation, so these additional measures become less important in terms of energy use.

## Using Landscape Elements to Provide Shade

Nothing can be much better at providing a cool shade in the summer than a great broad-leafed tree. In addition to shading the building from direct sun, trees have been found to reduce the temperature of air immediately around them by as much as 10°F (5°C) below the temperature of the surrounding air due to evaporation of moisture. A window shaded with a high tree or vine-covered trellis can have full shade in the summer, while enhancing the view and perhaps the ventilation. Trees and bushes can provide strategic shade from low east or west sun angles that are extremely difficult to shade architecturally. Similar to overhangs and other external shading devices, the impact of tree shading on solar heat gain through windows is less significant if high-performance, low-SHGC windows are used (Figure 5-17).

## Using Interior Shading Devices

Most homeowners use some form of interior window treatment, such as drapes, blinds, or shades, on their windows. In addition to their decorative aspects, drapes and curtains have been traditionally used by homeowners to control privacy and daylight (as noted earlier in this chapter), provide protection from overheating, and reduce the fading of fabrics.

To most effectively reduce solar heat gain, the drapery used to block the sunlight should have high reflectance and low transmittance. A densely woven fabric with a light color would achieve this objective. Drapes can reduce the solar heat gain coefficient of clear glass from 20 to 70 percent, depending upon the color and openness of the drapery fabric. The impact of drapery on the solar heat gain is proportionally lessened as the window is shaded by other methods, such as exterior shading or tinted glass. The main disadvantage of drapes and other interior devices as solar control measures is that once the solar energy has entered a window, a large proportion of the energy absorbed by the shading system will remain inside the house as heat gain. Interior devices are thus most effective when they are highly reflective, with minimum absorption of solar energy.

Blinds and shades primarily provide light and privacy control but they also can have an impact on controlling solar heat gain. They include horizontal venetian blinds, the newer miniblinds, vertical slatted blinds of various materials, a wide variety of pleated and honeycomb shades, and roll-down shades.

White- or silver-colored blinds, coupled with clear glass, have the greatest potential for reducing solar heat gains. Some manufacturers have offered window unit options that include miniblinds mounted inside sealed or unsealed insulating glass. The blinds, in the sealed dust-free environment, can be operated with a magnetic lever without breaking the air seal. Blinds in the unsealed glazing unit are protected as well, but can be easily removed if needed for cleaning or repair. These "between-glass shading devices" provide a lower shading coefficient than equivalent blinds mounted on the interior. They also provide additional insulating value to the double glass by reducing convective loops within the air space.

Like the other traditional strategies for reducing heat gain, interior shades and blinds can make an effective contribution when common single- or double-glazed windows are used. If high-performance glazing with a low solar heat gain coefficient is used (window 9 in Figure 5-18), the relative and absolute impact on cooling energy use is diminished. Figure 5-18 also illustrates the difference in performance related to shade color.

Unlike the other strategies to reduce heat gain, interior shades generally require consistent, active operation by the occupant. It is unlikely that anyone would operate all shades in a consistent, optimal pattern as they are assumed to be operated in Figure 5-18. It is possible to install motorized and automated shading systems, but these are costly and not yet in common use. By using high-performance glazing to provide the necessary solar control, there are two important benefits—there is less need for operating the shades, and the window is rarely covered, resulting in a clear view and daylight at all times. Of course, shades also provide privacy and darkness when desired, so they may be closed part of the day in any case, but the high-performance glazing means

Figure 5-18. Impact of interior shade operation on annual cooling season energy use in Phoenix, Arizona.

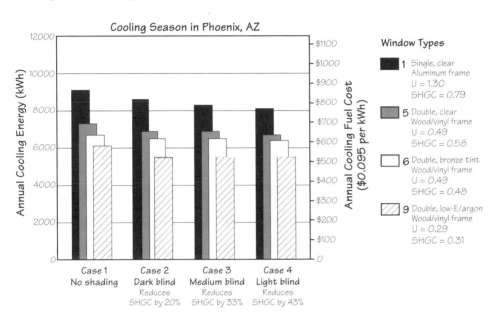

Note: The annual energy performance figures shown here are for a typical 1540 sq ft house. In all cases, the shade is assumed to be in place whenever the solar radiation is equal to or greater than 30 Btu/sq ft. U-factor and SHGC are for total window including frame. House and windows are described in Appendix A. kWh=kilowatt hours.

there is less need to operate them in a particular manner to make significant reductions in energy use. If your goal is to minimize cooling energy use, or you live in a house without air conditioning in a hot climate, then the combination of good shade management with low SHGC windows will be the best strategy.

## DECREASING WINTER HEAT LOSS

### Determining the Optimal Amount of Glazing

In the 1970s, when energy use was an emerging concern, the high-performance windows of today were not available. One of the obvious architectural design approaches was simple: to reduce heat loss, reduce window area. Furthermore, where windows were used, strategies such as using thermal shades had a major impact. As windows have improved considerably in the last twenty years, superwindows can equal or exceed the performance of insulated walls over a complete winter heating season. Consequently, the strategy of reducing window area to reduce energy use is no longer as significant if highly efficient windows are used.

As Figure 5-19 illustrates, total glazing area has a significant impact on heating energy use when single- and even double-glazed windows are used. With high-performance windows, however, the glazing area is not an important factor. In fact, with the superwindow (U = 0.15), the total heating energy use for a house in Madison, Wisconsin decreases slightly as the glazing area increases. This indicates that the benefit of more passive solar gain exceeds any losses from more glazing area. It should be noted, however, that cooling energy for the house with high-performance windows increases slightly with greater glazing area. Depending on the exact U-factor, SHGC, and climate, energy gains in the heating season may be offset by losses in the cooling season. However, by shifting the window area to preferred orientation and employing other cooling load reduction strategies for the windows as noted earlier, we can produce a design with large windows (25 percent glazing-to-floor-area ratio) whose total

---

### Guidelines for Decreasing Winter Heat Loss

Using clear single- or double-glazed windows:

- Concentrate glazing on southerly orientations to maximize solar benefits.

- Minimize window area to reduce winter heat loss.

- Use thermal shades or movable insulation over windows to reduce nighttime heat loss.

- Minimize windows facing prevailing winter winds and use landscape elements to block winter winds.

Using high-performance windows (low U-factor):

- Window area can be increased without a significant energy penalty. With superwindows, annual energy use can actually be reduced by adding more windows.

- Thermal shades and movable insulation are not necessary to improve energy performance.

Figure 5-19. Impact of window glazing area on annual heating and cooling season energy use in Madison, Wisconsin.

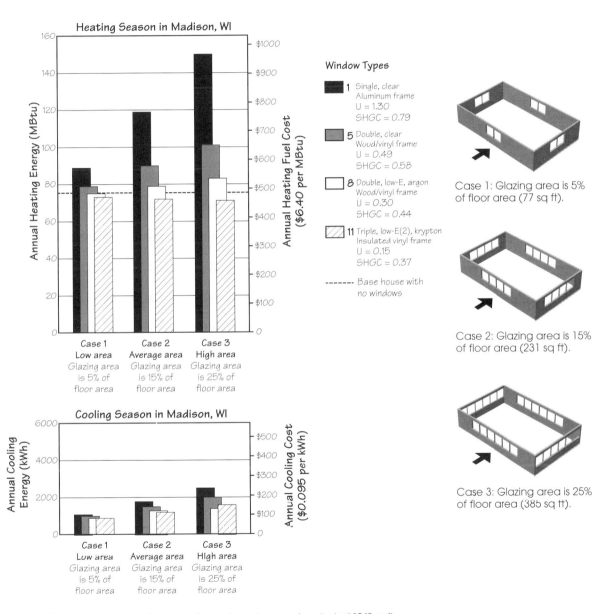

**Window Types**

1 Single, clear
Aluminum frame
U = 1.30
SHGC = 0.79

5 Double, clear
Wood/vinyl frame
U = 0.49
SHGC = 0.58

8 Double, low-E, argon
Wood/vinyl frame
U = 0.30
SHGC = 0.44

11 Triple, low-E(2), krypton
Insulated vinyl frame
U = 0.15
SHGC = 0.37

------- Base house with
no windows

Case 1: Glazing area is 5% of floor area (77 sq ft).

Case 2: Glazing area is 15% of floor area (231 sq ft).

Case 3: Glazing area is 25% of floor area (385 sq ft).

Note: The annual energy performance figures shown here are for a typical 1540 sq ft house. U-factor and SHGC are for total window including frame. House and windows are described in Appendix A. MBtu=millions of Btu. kWh=kilowatt hours.

energy use is less than the base case house with less glazing (15 percent glazing-to-floor-area ratio).

There are two important cautions in interpreting Figure 5-19. First, even though the performance of triple-glazed superwindows looks promising, there may be specific periods of cold weather and low sunshine during which superwindows will experience a loss and require heating. Second, although several manufacturers now offer commercially available superwindows, they are typically specialty products whose cost is somewhat higher than the more widespread double-glazed, low-E argon-filled windows.

## Using Shades and Insulated Shutters

Traditionally, drapes have provided a slight benefit in reducing heat loss and an important contribution in improving thermal comfort as radiant barriers. Because drapes are normally not tightly fit to the walls and floor, air can circulate around them, making them marginally effective in reducing conductive and convective heat loss. When a heavy drape is pulled across a cold window surface, the occupants of a room can feel considerably warmer because they are no longer radiating their warmth to the cold glass. Studies have suggested that in some cases room air temperatures can be lowered by as much as 5°F (3°C) while still maintaining the same comfort level if cold windows are covered with drapes. Convective air currents, however, can reduce comfort. When a window extends from floor to ceiling, a downdraft of chilled air can be created on the cold interior surface of the glass, which then flows out beneath the drapes and across the floor.

With energy concerns arising in the 1970s, movable insulating panels over windows became a popular strategy. Movable insulation has the advantages of providing high levels of insulation at night when most needed and allowing the greatest transparency for view and solar gain at other times. However, these panels were typically costly, required tight sealing to the window in order to work well, and needed consistent operation to be effective.

As with many other energy-efficient techniques developed in the 1970s, the value of thermal shades and movable insulation panels is reduced by high-performance windows (Figure 5-20). The relative contribution of movable insulation to overall home energy savings decreases as the insulating value of the windows themselves increases.

134

The insulating value of a standard double-glazed low-E window is about the same as a 3/4-inch (1.9 cm) layer of polystyrene insulation (R-3). A superwindow can approach the insulating value of 1 to 1-1/2 inches (2.5 to 3.8 cm) of polystyrene insulation (R-5 to R-7). The high-performance window is considerably more effective than movable insulation since the high-performance glazing requires no operation and is in place twenty-four hours a day. In addition, with no shades necessary, solar heat can enter and further reduce heating energy costs, particularly if a low-E glazing with a high SHGC is chosen.

Exterior rolling shutters with hollow or insulated slats have been popular in Europe and are available in the United States as well. They provide only modest additional insulating

Figure 5-20. Impact of insulating thermal shutters on annual heating season energy use in Madison, Wisconsin.

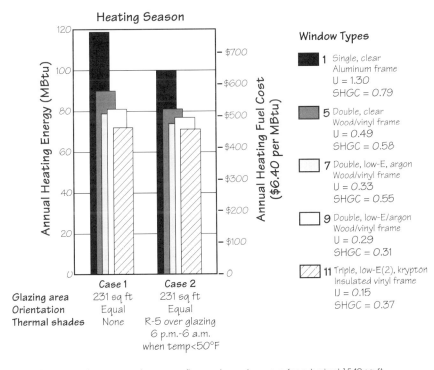

Note: The annual energy performance figures shown here are for a typical 1540 sq ft house. U-factor and SHGC are for total window including frame. House and windows are described in Appendix A. MBtu=millions of Btu.

value, but they also reduce wind effects, and provide solar control, privacy, and security. Most are operated manually, which means they are subject to the same inefficiencies of any occupant-operated device, but they can be motorized and controlled automatically.

## Minimizing the Effect of Winter Winds

Since infiltration is increased by wind pressure, windows directly facing prevailing winter winds will experience more air leakage than those oriented away from the winds. In many northerly climates, this means minimizing windows toward the north and west, although precise wind patterns depend on many factors, including topography, landscaping, and other site-specific characteristics. Usually this does not conflict with placing windows toward the south for beneficial solar gain.

When windows do face toward prevailing winter winds, landscape elements such as earth berms, walls, and thick rows of evergreen trees or shrubs can be effective wind-breaks. This protection has an impact not only on window infiltration but also on air leakage through the entire building envelope. As with other traditional energy-saving strategies, the beneficial impact of wind protection by external elements is reduced as the building and its windows become more airtight.

## PROVIDING WINTER SOLAR HEAT GAIN

Providing winter solar heat gain has a long history in architectural design, and it has been aggressively promoted in the United States as a specific energy efficiency strategy since the 1970s. Generally, passive systems have been based on the simple principle that solar gain will be maximized by placing more glazing toward the south and less glazing in other directions. This implies arranging the floor plan along an east-west axis so that more rooms have southern exposure. In addition, the amount of south-facing glazing in each room can be increased to optimize solar gain. There are generally two constraints to the use of increased south-facing window area. Each additional unit of glass area provides diminishing benefits but continues to increase nighttime heat losses, so there is an optimal area. In addition, in order to be useful, solar heat admitted through windows must be stored in thermal mass within the house, to be released later in the afternoon and at night. An inadequate ratio of mass to glass area will result in overheating during winter days and less useful solar gain than expected. Finally, the large solar aperture must be carefully protected during the cooling season to avoid increased air-conditioning costs. A number of houses have been designed that successfully use this south-facing glass combined with mass strategy to reduce heating and cooling costs. But this approach has not been adopted widely because it typically involves several departures from traditional design and construction techniques.

The improved insulating value and solar control characteristics of high-performance glazing have modified some traditional energy design assumptions described previously. With passive solar design, the availability of high-performance glazings requires some rethinking of traditional assumptions as well, but it also presents new opportunities to enhance passive design. For example, east-, west-, and even north-facing windows can become passive solar collectors, meaning that their useful solar contribution exceeds their losses (although they will never provide as much winter solar gain as a south-facing window). This provides more architec-

---

**Guidelines for Providing Winter Heat Gain**

- Locate the house so that other buildings or landscape elements do not obstruct solar gain. Use deciduous plantings to permit solar gain during winter but block summer sun.

- For all types of glazing, face windows to the south to permit maximum solar gain in winter. However, when using high-performance windows, orientation has less impact on energy performance in a typical house without significant thermal mass.

- To maximize useful solar gain, use glazing with high solar heat gain coefficients and suitable thermal mass.

- When using double glazing or high-performance windows, south-facing window area can be increased with suitable thermal mass to provide lower winter heating energy use.

tural design freedom and should result in greater acceptance of these design approaches.

## Providing Solar Access

In order to receive solar radiation in winter, the house must be located so that it is not in the shadow of other buildings or landscape elements. If possible, locating the building on the north end of the site (in the northern hemisphere) provides a greater assurance of future solar access. Of course, deciduous trees can be located within these limits.

As discussed in the previous section, overhangs on south-facing walls can be sized to permit low-angled winter sun to penetrate windows while still blocking much of the higher-angled summer sun. Most other external and internal shading devices such as awnings, shutters, or shades are operable and can be adjusted seasonally or daily to maximize solar gain in winter.

The advantage to using deciduous vegetation for shading is that its cycle follows the local climate. Trees leaf out in the spring when the weather warms up, just in time to provide shading. They drop their leaves when the cold snap hits, earlier in northern climates than in southern, allowing the sun's heat to penetrate and warm the home. This modulation of shading from trees, shrubs, and vines is wonderfully automatic, but it does require a commitment to landscape maintenance by the homeowner. It should be noted that the branches of a deciduous tree without its leaves still block some sunlight.

## Orienting Windows to Maximize Solar Gain

It is generally accepted that simply orienting the majority of windows to the south in a heating-dominated climate will result in greater solar gain and less heating energy use. However, it is not always feasible to do so. In these cases east and west windows might be useful as secondary solar collectors using high-performance windows. To test this assumption, the same typical house was used to simulate energy use for five different orientations in a cold climate (Madison, Wisconsin). The results shown in Figure 5-22 indicate that, as expected, south-oriented windows perform best, and there is a difference between orientations, particularly with single and double glazings. However, when higher-performance windows with

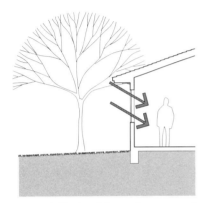

Figure 5-21. Deciduous trees permit solar gain during winter.

Figure 5-22. Impact of window orientation on annual heating season energy use in Madison, Wisconsin.

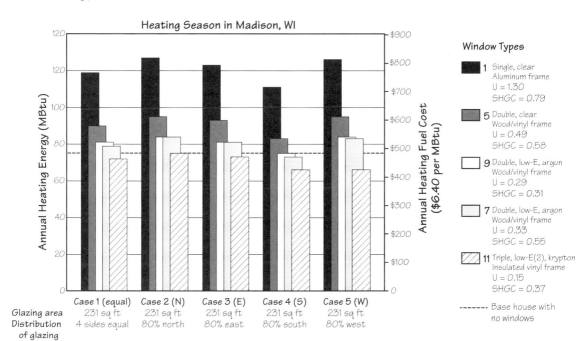

Note: The annual energy performance figures shown here are for a typical 1540 sq ft house. U-factor and SHGC are for total window including frame. House and windows are described in Appendix A. MBtu=millions of Btu.

lower U-factors are used, the difference between orientations becomes less.

It is useful to compare window 7, a low-E window with a high solar gain coating (SHGC = 0.55), and window 9, which also has a low-E coating but with lower solar gain (SHGC = 0.31), since it was designed to reduce summer cooling loads as well. Window 7 is better on the south by a small amount, while the windows are similar on the other orientations. Note that window 9 does have a slightly lower U-factor, which helps it in all orientations. Note also that for the southerly orientation, these two windows reduce the house heating to the level of a house without any windows. Window 11, the low U-factor superwindow (U = 0.15), is the best performer in all orientations, even though its SHGC is not high because of all the glazings and coatings used.

Another observation that can be made in Figure 5-22 is that the double-glazed windows with low-E on the south have about the same performance as the superwindows on the north. Thus, if you have flexibility with siting and house layout, lower energy consumption can be achieved using less costly low-E windows facing in southerly orientations. If other orientations are necessary, equivalent performance can then be obtained by switching to the higher-performance (and more costly) window designs. Some of the economic differences involved are relatively small, as can be seen from the data in the figure. Note, as well, that any attempt to optimize orientation must account for the cooling-related energy use discussed earlier. Fortunately, south-facing windows are the easiest to shade in summer, as discussed previously in this chapter.

The fact that the two low-E windows with very different SHGCs performed similarly suggests that the solar gain is not being used optimally in this house. In fact, the house has relatively low thermal mass (just the intrinsic mass of its standard lightweight construction and furnishings), which might be increased to improve the performance of south facing windows, as discussed in the next section.

## Passive Solar Design

To make best use of the sunlight available through south glazing, it is important to provide enough thermal mass so that the sun's energy is absorbed in the mass and then transferred to the room air later at night when the temperature falls and heating is called for. A classic mistake in early passive solar designs was not having adequate thermal mass, resulting in overheating during the day and inadequate heat stored to meet the night heating requirement.

In Figure 5-23 the analysis in Figure 5-22 is extended to examine the effect of added mass when extensive south glazing is used. Case 1 is the conventional house with equal glazing on all four sides (the same as case 1 in Figure 5-22). In case 2 all the glazing is oriented to the south and an overhang is added for summer sun protection. In case 3 interior thermal mass equivalent to an exposed 4-inch (10 cm) concrete slab is added throughout the house. Comparing case 2 and 3 to see the effect of mass, we see two distinct situations with different glazings. For the two glazings with lower SHGC (windows 9

Figure 5-23. Impact of passive solar strategies on annual heating and cooling season energy use in Madison, Wisconsin.

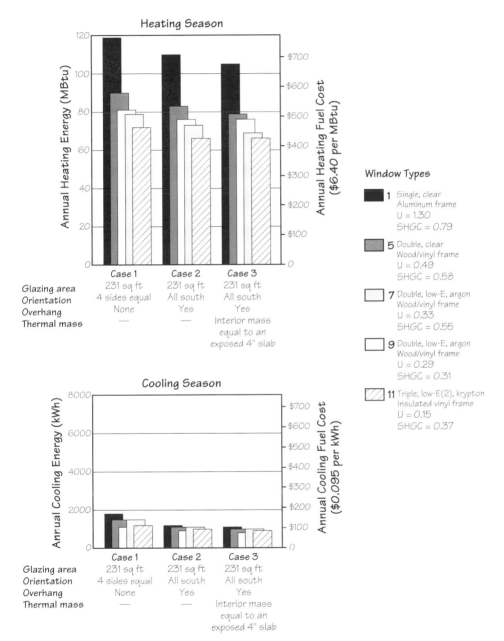

Note: The annual energy performance figures shown here are for a typical 1540 sq ft house. U-factor and SHGC are for total window including frame. House and windows are described in Appendix A. MBtu=millions of Btu. kWh=kilowatt hours.

and 11), there is no change with the addition of mass, suggesting that whatever solar gain enters the house is already effectively used to offset daytime and early evening losses and there is not enough excess gain to be stored and used later at night. However, in the case of the two high solar gain glazings (windows 5 and 7), there are noticeable benefits from adding mass. If high solar gain glazings and adequately-sized thermal mass are used, then further increasing the south-facing windows in size would reduce heating energy use. In very cold climates, the south glazing area should be as much as 35 percent of the floor area (it is 15 percent here) to provide a very significant reduction in home energy use. With glazings of this size it is essential that summer sun control is carefully designed and that thermal mass is carefully sized and located. Investigate passive solar design if you want to explore these design options further.

# CHAPTER 6

# Window Selection Considerations

Selecting a window involves many considerations related to appearance, function, energy performance, and cost. Seeking the desired appearance of a window in terms of both the exterior facade and the interior design is often the starting point. Appearance includes the size, shape, style, and materials of the window. Once these aesthetic characteristics are

## Window Selection Considerations

**Appearance**
- Size and shape
- Style
- Frame materials
- Glass color and clarity

**Energy performance**
- Basic energy-related properties
- Annual heating and cooling season performance
- Peak load impacts
- Long-term ability to maintain energy performance

**Function**
- Daylight
- Glare control
- Fading
- Thermal comfort
- Resistance to condensation
- Ventilation/operating type
- Sound control
- Maintenance
- Durability (warranty)

**Cost**
- Initial cost of window units and installation
- Cost of interior and exterior window treatments
- Cost of maintenance
- Frequency of replacement
- Resale value
- Initial cost of heating and cooling system
- Annual cost of heating and cooling energy

identified, the basic functional aspects of the window are addressed; these include daylight transmission, ventilation operation, and condensation resistance. Next, energy performance represents a set of decisions that can have a significant impact on operating cost as well as comfort. Finally, cost in general is an important factor in any selection process. Cost, however, does not mean just the initial purchase price. Many interrelated costs occur throughout the life of a window.

Often, selecting a window is based on only some of these many considerations. In the simplest case, a decision to purchase a window can be based on the basic style, operating type, and initial cost. The problem with this approach is that there is no certainty of performance. Often, the energy-related aspects of the window are not factored into the decision. While this approach may save money on the initial purchase, it may result in considerably more expense, both directly and indirectly, over the life of the window. For example, you may be unable to use a room that is consistently hard to keep warm in winter because cheap, inefficient windows were selected.

New technologies that improve window performance and reduce wasteful energy use could improve local air quality and reduce greenhouse gas emissions. Thus, choosing energy-efficient windows can have significance for the society as a whole as well as for the individual.

The purpose of this chapter is to describe the window selection considerations and indicate how each one plays into a purchasing decision. At the end, this approach to window selection is summarized in a checklist.

## APPEARANCE

| Appearance |
| --- |
| • Size and shape |
| • Style |
| • Frame materials |
| • Glass color and clarity |

The way a window or skylight looks can sometimes override all other technical and cost considerations. After all, a primary basis for selecting a window is to help achieve a certain exterior and interior design concept. The appearance of the window frame—its style, materials, and use of mullions—is perhaps one of the most important considerations in selecting any window. In most cases, the desired size and shape of a window can be found in numerous product lines. In a few cases, the need for an unusual shape, such as an arched window, may limit the buyer to certain manufacturers.

The color and finish of the interior frame materials (wood, vinyl, or metal) is a major selection criterion for many designers and homeowners: vinyl and metal exterior frame material or cladding may not be paintable. Fiberglass and wood, of course, allow for choice of paint color. Choosing window size and shape and frame type are subjective decisions–they depend on the house design and the personal preferences of the designer or homeowner.

The appearance of the glazing—its visual clarity—is another issue. Most new good-quality glass has minimal optical distortion, although differences in temperature and pressure can cause sealed insulated glass units to bow slightly. Tempered or heat-strengthened glass can show more distortion

Figure 6-1. The style and overall appearance of windows are primary selection considerations. (Photo: Summit Window & Patio Door.)

than standard window glass, particularly at the edges. Another visual characteristic is haze, the degree to which the glass itself diffuses light and softens the view of objects beyond. With modern glass technology, most consumers need not consider visual clarity unless a window obviously appears distorted or hazy.

Glass color is a matter of importance to some homeowners. Not only does tinted glass diminish the amount of light, but the light that passes through it appears colored—either gray, bronze, green, or blue-green—and it can change both the appearance of objects indoors seen under the colored light and objects viewed through the glass. With new glazing products such as spectrally selective low-E coatings, solar heat gain can be reduced without the appearance of tinted glass. However, some low-E coatings impart a subtle color shift, especially when viewed at an angle.

These are aesthetic characteristics worth considering. The best way to assess these effects is to observe sample glazings under different light conditions and, if possible, visit a house that uses the glazings being considered.

## FUNCTION

The functional issues in window selection include a variety of practical considerations, such as providing light, view, and fresh air while maintaining visual and thermal comfort. A window unit must also meet a number of technical performance standards and tests to ensure that it performs adequately. Windows available on the market must meet numerous standards and building regulations, so many of these basic technical characteristics are not of direct concern to the consumer selecting a window. These issues include keeping out water and having sufficient structural strength in the frame and glazing. In special cases, such as a building in hurricane-prone area, building codes may require glazing of a certain strength and type to resist breakage during storm conditions. A few technical considerations, however, are important to many window purchasers: condensation resistance, sound control, maintenance requirements, and overall durability of the unit. This section discusses selecting windows according to their basic functions as well as other key technical and practical concerns.

**Function**
- Daylight
- Glare control
- Fading
- Thermal comfort
- Resistance to condensation
- Ventilation/operating type
- Sound control
- Maintenance
- Durability

146

Figure 6-2. The desire for daylight is one major influence on the design and selection of windows. (Photo: Velux-America Inc.)

## Daylight

The amount of daylight entering a window depends on a number of external factors including the direction the window faces, whether trees or buildings block sunlight, and the presence of building elements, such as overhangs, awnings, or other shading devices. Once the daylight reaches the window, the glazing itself influences how much light is transmitted to the interior. The percentage of visible light that is transmitted through the window is indicated by the visible light transmittance (VT). The higher the VT, the more light is transmitted. Note that the VT rating appearing on an NFRC label includes the combined effect of the glazing and the frame, which blocks all light. In Figure 6-3, five of the six glazings have a total window VT of 0.40 or above, which is equivalent to a transmittance through the glazing alone of about 0.60. If you compare NFRC ratings to other light transmittance data for glass only, be sure to adjust for the effect of the frame.

In many cases, lowering the solar heat gain coefficient is a primary goal. Typically, this means that the amount of light is

Figure 6-3. Basic characteristics of typical window glazings.

| Window description | U-factor (Btu/hr-ft²-°F) | Solar Heat Gain Coefficient | Visible Transmittance | Light-to-solar-gain ratio |
|---|---|---|---|---|
| 5 Double glass | 0.49 | 0.58 | 0.57 | 0.98 |
| 6 Double glass, bronze | 0.49 | 0.48 | 0.43 | 0.89 |
| 8 Double glass Low-E, argon | 0.30 | 0.44 | 0.58 | 1.31 |
| 9 Double glass Selective low-E , argon | 0.29 | 0.31 | 0.51 | 1.65 |
| 10 Double glass Selective low-E, argon | 0.31 | 0.26 | 0.31 | 1.19 |
| 11 Triple glass Low-E (2), krypton | 0.15 | 0.37 | 0.48 | 1.29 |

All values in this table are for the whole window and include the effects of the frame.
See Appendix A for detailed descriptions of the windows.

reduced as well, as is evident when clear and tinted glazings are compared. Figure 6-3 (windows 5 and 6) shows that using tinted glass lowers the SHGC somewhat but also lowers visible light transmittance to a greater degree. Windows 8 and 9 utilize special low-E coatings, which provide better solar heat gain reduction than tinted glass, with a minimal loss of visible light. Low-E/spectrally selective coatings can be formulated to achieve a variety of performance goals. Window 10 illustrates a coating designed to minimize SHGC as much as possible while still maintaining a reasonable, although reduced, amount of light. Window 11 is a three-layer superwindow with good SHGC and VT characteristics.

In window selection, the designer or homeowner must determine the relative importance of solar control and daylight, then make a glazing choice that emphasizes either one or both of these. Not all manufacturers, however, offer all types of glazings in all of their product lines. Information on SHGC and VT is found in manufacturers' catalogs and on the NFRC label. Figure 6-3 demonstrates that it is possible to select a window with a low SHGC without sacrificing daylight transmittance. Windows with a high light-to-solar-gain (LSG) ratio meet this requirement.

## Glare Control

Although daylight is generally desirable, too much direct light can result in uncomfortable glare. Providing good visual comfort should be a priority in rooms that are used for demanding visual tasks, such as home offices. If a window that reduces visible light transmittance is chosen, then the potential for glare will be reduced as well. However, the goal is often to provide as much daylight as possible while controlling glare problems. This can be handled in the architectural design of the space as well as by the use of diffusing curtains or shades. Techniques for managing daylight and controlling glare are described in Chapter 5. Note that the NFRC value (VT) accounts for the effect of the frame as well as the glazing in blocking light.

## Fading

Many organic materials, such as carpet, fabrics, paper, artwork, paints, and wood may fade upon exposure to sunlight. The potential for damage largely depends on: (1) the nature of the materials and dyes in the materials; (2) the type of radiation (UV, daylight) to which they are subjected; (3) the intensity of radiation; and (4) the duration of exposure. Window selection can influence the type and intensity of transmitted radiation. The most harmful radiation in sunlight is the ultraviolet (UV) rays, which are the most energetic and thus most likely to break chemical bonds, leading to fading and degradation. However, the blue end of the visible spectrum can also contribute to fading. The amount of harmful UV and visible radiation depends on a variety of climatic factors, such as water vapor levels and pollution in the atmosphere. Glass blocks all UV radiation below 300 nm, but transmits UV from 300–380 nm. Coatings on glass can reduce the UV transmitted by up to 75 percent. UV absorbers can be incorporated into thin plastic films in multilayer windows or as an interlayer in laminated glass. In both cases, the UV transmission can be reduced to less than 1 percent. However, it is important to note that the remaining visible light that is transmitted can still cause serious fading in some materials. Using low-E coated glass or windows incorporating plastic layers rather than clear uncoated glass will reduce fading for many modern interior furnishings. However, none of these strategies will provide complete protection from fading. Place valuable artifacts that are highly susceptible to fading in hallways or other spaces without exposure to substantial window area, or in rooms that have appropriate drapes and shades to substantially reduce daylight transmittance. Museum conservators limit light levels to less than 150 lux, or only 1/700 the level of bright sunlight, to protect artwork.

## Thermal Comfort

The degree to which a window provides thermal comfort is an important but difficult-to-quantify consideration in selecting windows. In winter, thermal comfort near a window is provided by minimizing cold air leakage, and by maximizing the temperature of the glass itself (by using glass with a low U-factor). A window with a lower glass temperature feels colder because more heat is radiated from a person's body to the window. Cold glass can also create uncomfortable drafts—air next to the window is cooled and drops to the floor. It is then replaced by warmer air from the ceiling, which in turn is cooled. This sets up an air movement pattern that feels drafty and accelerates heat loss. Selecting a window with a lower U-factor will result in a higher interior window temperature in winter and thus greater comfort. A relatively tight window will improve comfort by reducing cold air leakage.

In summer, comfort is based on reducing solar heat gain through the windows and by reducing the glass temperature. By selecting a window with a lower solar heat gain coefficient, the direct radiation striking a person will be reduced. By selecting glazings that reduce solar heat gain by reflection instead of absorption, the interior window temperature is lowered and comfort improved.

In a climate where winter comfort is more important than reducing solar gain in summer, a window with a low U-factor but a relatively higher SHGC is a good choice (see window 8 in Figure 6-3). Feeling the warmth of directly transmitted sunlight in winter can be important to thermal comfort.

A window with poor energy performance characteristics will also be an uncomfortable window. A window with good energy performance characteristics will generally provide greater thermal comfort than a poorer energy performer. Paying more for better-performing windows is often viewed only in relationship to fuel cost savings; however, thermal comfort is a clear benefit and should be considered an addi-

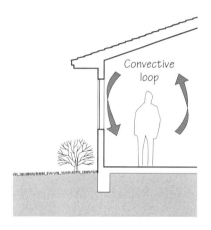

Figure 6-4. Drafts occur as air next to the window is cooled and drops to the floor. It is then replaced by warmer air from the ceiling creating a convective loop.

tional value worth paying for when selecting windows. There is little point in spending money to improve or add a room that is unusable at times because of thermal comfort problems.

## Resistance to Condensation

Condensation has been a persistent and often misunderstood problem associated with windows. Single-glazed windows characteristically suffer from water condensation and the formation of frost on the inside surface of the glass in winter. The surface temperature of the glass drops below either the dew point or frost point of inside room air. Frost patterns on the windows were often one of the first signs of approaching winter.

Insulated windows create a warmer interior glass surface, reducing frost and condensation. Insulated "superwindows" with warm edge technology and insulating frames have such a warm interior surface that condensation on any interior surfaces may be eliminated if humidity levels are maintained at reasonable levels. Graphs such as the one shown in Figure 6-5 must be used with caution since they show results for center glass only. As the thermograms in Chapter 3 illustrate, the temperature at the glass edge or the frame may be lower than the center of glass in windows that use two- and three-layer low-E, gas-filled glazings. Condensation may form at the coldest locations, such as the lower corners or edges of an insulated window even when the center of glass is above the limit for condensation.

Excessive condensation can contribute to the growth of mold or mildew, damage painted surfaces, and eventually rot wood trim. Since the interior humidity level is a contributing factor, reducing interior humidity is an important component of controlling condensation. This is done by first removing humidity at its source—with vent fans in kitchens and bathrooms and limited use of humidifiers. Then, the remaining humid air can be diluted with drier outside air. In tight, newer houses an air-to-air heat exchanger is often required for ventilation and accomplishes humidity removal as well. Finally, dehumidification can be used if necessary.

Condensation can also be a problem on the interior surfaces of window frames. Metal frames, in particular, conduct heat very quickly, and will "sweat" or frost up in cold weather. Solving this condensation problem was a major motivation

Figure 6-5. Condensation potential on glazing (center of glass) at various outdoor temperature and indoor relative humidity conditions. Condensation can occur at any points that fall on or above the curves. (Note: All air spaces are 1/2 inch; all coatings are e = 0.10.)

Example 1: At 20°F (-7°C) outside temperature, condensation will form on the inner surface of double glazing any time the indoor relative humidity is 52 percent or higher. It will form at an indoor relative humidity of 70 percent or higher if a double-pane window with low-E and argon is used.

Example 2: In a cold climate where winter night temperatures drop to -10°F (-23°C), we want to maintain 65% humidity without condensation. A double-glazed window with low-E and argon will show condensation at 57% relative humidity, so the triple glazing with two low-E coatings and argon is needed to prevent condensation.

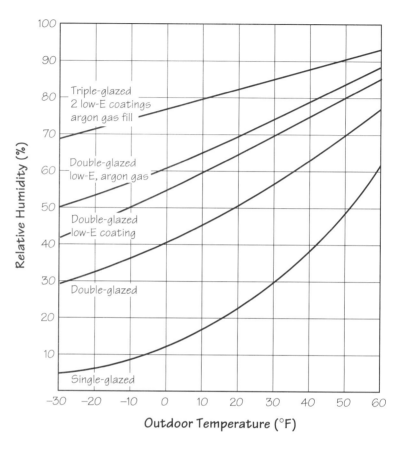

for the development of thermal breaks for aluminum windows (see Chapter 4).

At this time there is no easy-to-use rating for condensation resistance that addresses these detailed edge and frame effects. However, AAMA (American Architectural Manufacturers Association) has developed the Condensation Resistance Factor (CRF). Temperature measurements are used to determine an average surface temperature for the glass and for the frame, from which the CRFs are calculated. The lower of the two is used to determine the window CRF. A CRF above 35 is considered a thermalized window. Higher CRFs are desirable in more severe climates.

The NFRC is developing a new condensation resistance rating that will appear on the NFRC label. Until such a rating is available, in comparing windows of similar design, lower

U-factor generally indicates greater condensation resistance.

Infiltration effects can also combine with condensation to create problems. If a path exists for warm, moisture-laden air to move through or around the window frames, the moisture will condense wherever it hits its dew-point temperature, often inside the building wall. This condensation can contribute to the growth of mold on the frame or wall, causing health problems for some people, and it encourages the rotting or rusting of window frames. Frames must be properly sealed within the wall opening to prevent this potential problem.

Condensation can cause problems in skylights and roof windows as well as typical windows. "Leaky" skylights are frequently misdiagnosed. What are perceived to be drops of water from a leak are more often drops of water condensing on the cold skylight surfaces. Use of more highly insulating glazing and careful detailing of the skylight frame can solve this problem. In many systems, a small "gutter" is formed into the interior frame of the skylight where condensate can collect harmlessly until it evaporates back into the room air.

## Outdoor Condensation

Under some climate conditions, condensation may occur on the exterior glass surface of a window. This is more likely to occur on higher-performance windows with low-E coatings or films, and low-conductance gas fills that create very low U-factors. For exterior condensation to occur, the glass temperature must be below the outdoor dewpoint temperature. This is most likely to happen when there is a clear night sky, still air, and high relative humidity, in addition to the right temperature conditions. Like other dew formed at night, exterior window condensation will disappear as surfaces are warmed by the sun. It is the excellent thermal performance of well-insulated glazing that creates the condition where the outer glass surface can be cold enough to cause condensation to form.

## Condensation Between Glazings

A more annoying problem can arise with double-pane windows: condensation between the panes. Moisture can migrate into the space between the panes of glass and condense on the colder surface of the exterior pane. This condensation is annoying not only because it clouds the view, but because it may mean that the glazing unit must be replaced (if it is a sealed insulating glass unit). In a nonsealed unit, simpler remedies may correct the situation.

Factory-sealed insulated glass utilizes a permanent seal to prevent the introduction of moisture. The void may be filled with air or dry gases, such as argon. A desiccant material in the edge spacer between the panes is used to absorb any residual moisture in the unit when it is fabricated or any small amount that might migrate into the unit over many years. These windows will fog up when moisture leaking into the air space through the seals overwhelms the ability of the desiccant to absorb it. This could happen early in the window's life (the first few years) if there is a manufacturing defect, or many decades later because of diffusion through the sealant. Quality control in manufacture, sealant selection, window design, and even installation all can influence the rate of failure. Once a sealed window unit fails, it is not generally possible to fix it, and the sealed unit must be replaced. Moisture in the unit is also likely to reduce the effectiveness of low-E coatings and

Figure 6-6. Providing ventilation is a basic window function that influences design and selection. (Photo: Velux-America Inc.)

suggests that gas fills may be leaking out. Most manufacturers offer a warranty against sealed-glass failure which varies from a limited period to the lifetime of the window.

When condensation occurs between glazings in a nonsealed unit, there are several possible remedies. Most manufacturers who offer nonsealed double glazing include a small tube connecting the air space to the outside air, which tends to be dry during winter months. Check to be sure that the inner glazing seals tightly to the sash, and clear the air tube if it has become obstructed. In some cases, reducing interior room humidity levels may help alleviate the problem.

## Ventilation

Another functional consideration related to windows is ventilation. The ventilation capability of a window unit is based on the opening size and the frame operating type. Hinged windows (casements, awnings, and hoppers) can be opened so that the full window area provides ventilation. Sliding windows (double hung and horizontal sliders) only can be opened to provide ventilation through about half of the total window area. The ability of a window to provide good ventilation is not just a function of the window unit, but is more strongly influenced by the placement of the windows in relationship to breezes and the other windows, as well as the house layout. Chapter 5 addresses ventilation strategies in window placement. Ventilation is a concern in selecting windows mainly in terms of the opening size and frame operating type. Issues of window placement are part of the design, not the selection, process.

## Sound Control

The acoustic properties of windows are tied very closely to technologies for infiltration control and heat loss control. Air leakage, whether through a purposely open window or cracks around the frame, also allows sound to "leak" into the building. Sealing these cracks directly controls the sound penetration through the window.

The thicknesses of the glass and of any air spaces between multiple layers of glass also affect the acoustic properties. Sometimes windows are given a rating called *Sound Transmission Class* (STC), measured in decibels (dB). The STC rating was an interim rating procedure developed for interior partitions where noise at the frequency range of speech and office equipment is a concern. The proper rating for exterior walls and windows is *Outdoor-Indoor Transmission Class* (OITC), based on frequencies for aircraft, traffic, and trains.

A side benefit to houses with tight, insulated windows is that they are dramatically quieter than older houses with leaky, single-pane windows. This reduction in noise transmission from the outside to the inside can become a major benefit for homes located in noisy locations, such as those near a highway or airport. Some windows are specially designed to minimize sound transmission. Consult window suppliers for more information about particular products.

## Maintenance

Older windows with wood exposed on the exterior require periodic painting and replacement of the glazing compound that holds the glass in place. Newer windows made with vinyl or metal frames or cladding over the wood frame require no exterior maintenance. If the exterior frame material is wood, then the cost and effort of periodic painting should be considered in selecting a window. However, many new wood windows have a very durable exterior finish, applied in the factory, which is far better than traditional paint finishes.

## Durability

A durable window unit does not deteriorate rapidly and lasts a long time. This means that quality and function last over the life of the window unit. Since many windows appear similar in most respects at first glance, assessing the long-term performance can be difficult. There is no rating or absolute guarantee of the durability of a window. Just as in making any other important choices, the designer or homeowner must study the design and workmanship of the window, and then rely on the recommendations from others who have used similar products and the manufacturers' reputation. The advice of experienced architects and builders can be helpful to the homeowner. As with other products, warranties can be an indicator of the reliability of the window and its manufacturer. AAMA has developed a life-cycle testing procedure to assist in product evaluation. ASTM is in the process of developing a similar standard.

# ENERGY PERFORMANCE CONSIDERATIONS

One important practical reason to select energy efficient windows is to reduce the annual cost of heating and cooling your home. This makes good economic sense for most building owners and it also contributes to national and global efforts to reduce the environmental impacts of non-renewable energy use. It can be a relatively painless and even profitable way (see page 177) for every family to help improve the environment in which we live. In order to select a window which will lower heating and cooling costs, you first need to estimate how much energy the furnace and air conditioner will consume. This is influenced not only by the window properties as you would expect, but by a series of other factors including the house location and microclimate, house characteristics, occupant use patterns, and cost of energy.

Getting an accurate estimate of the annual energy consumption can take a little analysis. As explained below, this can best be done today with some simplified computer tools. But it is not always important to get an accurate quantitative analysis of energy and cost savings—frequently a comparative analysis or ranking is sufficient to guide your choice. For example, you may already have narrowed the decision to two or three window options and just want to pick the one with lower heating energy use. In this case a more simplified set of guidelines may suffice and the new heating and cooling energy ratings being developed by NFRC will be most appropriate.

To evaluate windows with respect to energy performance, various types of information and tools are available:

1. Evaluate the window based on its energy-related properties applied to your climate.

2. Use an annual energy performance rating system to evaluate heating and cooling energy use.

3. Use a computer program to compare energy use and utility costs.

Each of these approaches is described in more detail in the remainder of this section. Selecting a window based on energy performance may involve two additional considerations—the impact on peak heating and cooling loads, and the long-term ability of the window unit to maintain its energy performance characteristics. These additional energy issues are discussed

> **Energy performance**
> - Basic energy-related properties
> - Annual heating and cooling season performance
> - Peak load impacts
> - Long-term ability to maintain energy performance

157

at the end of this section along with the role of codes and standards in improving window energy efficiency.

In addition to quantitative issues such as the actual cost of heating fuel or electricity for cooling, energy performance characteristics are linked to other less measurable issues such as thermal comfort and condensation resistance, as noted earlier in the chapter. Choosing a better-performing window to save on fuel costs will also improve comfort and performance in these other areas.

## Using the Basic Energy-Related Properties

The three key properties are U-factor, solar heat gain coefficient (SHGC), and air leakage rating. Visible transmittance (VT) is another property used in comparing windows. These are defined and described in detail in Chapter 2 and are the first properties to appear on NFRC window labels (U-factor was introduced first, followed by SHGC and VT; air leakage ratings will be added soon).

Figure 6-7 indicates some guidelines for using the basic energy properties in choosing a window. Note that these guidelines are different for distinct climate regions. Until there is a reliable annual performance rating in place or unless you are using a computer program, these properties are the main basis for making energy performance decisions.

## Using an Annual Energy Performance Rating

Even though there are code minimums, guidelines, and recommended levels for the basic properties found on the NFRC label, they do not, in themselves, give the consumer a clear indication of the actual impact on energy costs. To accurately determine annual energy performance and cost, they must be calculated using a sophisticated computer program that takes into account the properties of the windows you are comparing, and a detailed description of your house, the climate, and the way in which you will operate the house. Unfortunately, these computerized tools are not always accessible to designers, builders, and homeowners to use in making a window purchasing decision, and even if they are, they may be too time consuming to use.

To overcome the limitations of requiring detailed computer simulations for each situation, the window industry is developing a simplified annual energy rating system for win-

Figure 6-7. Energy-related properties of windows.

| | Description | Heating Climate | Mixed Climate | Cooling Climate |
|---|---|---|---|---|
| **Heat Flow (U-value)**  | The rate of heat transfer is indicated in terms of the U-value (U-factor) of a window assembly. The insulating value is indicated by the R-value, which is the inverse of the U-value. The lower the U-value, the greater a window's resistance to heat flow and the better its insulating value. | PRIMARY FACTOR: A low U-value is the most important window property in cold climates. | PRIMARY FACTOR: A low U-value is one important window property in mixed climates. | A low U-value is helpful during hot days or whenever heating is needed, but it is less important than SHGC in warm climates. |
| **Solar Heat Gain (SHGC)** 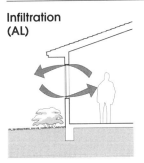 | The SHGC is the fraction of incident solar radiation admitted through a window, both directly transmitted, and absorbed and subsequently released inward. SHGC is expressed as a number between 0 and 1. The lower a window's SHGC, the less solar heat it transmits. | A high SHGC increases passive solar gain for heating, but reduces cooling season performance. A low SHGC improves cooling season performance, but reduces passive solar heating. | PRIMARY FACTOR: A low SHGC is one important window property in mixed climates. | PRIMARY FACTOR: A low SHGC is the most important window property in warm climates. |
| **Infiltration (AL)**  | Heat loss and gain occur by infiltration through leaks in the window assembly. The air leakage rating (AL) is expressed as cubic feet of air passing through an equivalent square foot of window area. The lower the AL, the less air will pass through leaks in the window assembly. | The air leakage rating (AL) is an important window property in cold climates. Air leakage should not exceed 0.56 cfm/sq ft. | The air leakage rating (AL) is an important window property in mixed climates. Air leakage should not exceed 0.56 cfm/sq ft. | The air leakage rating (AL) is generally less important in warm climates. However, infiltration can contribute to excessive summer cooling loads by introducing humid outdoor air. |
| **Daylight (VT)** | The visible transmittance (VT) is an optical property that indicates the amount of visible light transmitted through the glass. VT is expressed as a number between 0 and 1. The higher the VT, the more daylight is transmitted. | A high VT is desirable to maximize daylight and view. | A high VT is desirable to maximize daylight and view. | A high VT is desirable to maximize daylight and view, but this must be balanced against the need to control solar gain and glare in hot climates. |

Note: See pages 164–168 for maps of heating, mixed and cooling climate regions in the United States

dows as well as a companion computer-based approach. This annual energy rating, currently being refined by the U.S. Department of Energy and window industry researchers in cooperation with the National Fenestration Rating Council, will be adopted as part of the official rating system of the NFRC.

### Development of the NFRC Annual Energy Rating

This section describes the process by which NFRC is developing its annual energy ratings. In order to understand the impact of window selection on annual energy use, researchers performed thousands of computer simulations (using DOE 2.1E) of houses with a wide range of characteristics in many U.S. climates. These simulations revealed that while the actual energy use varied as house characteristics were changed, the relative impact of changing window types was often unchanged. For example, if changing from one window to another resulted in a 20 percent annual heating energy savings, the savings remained at 20 percent even if many of the other house characteristics, such as insulation levels, were changed. As shown in Figure 6-8, the relative performance of four different windows is similar in a wide range of climates (with a few exceptions that do not have great significance).

With this as background, NFRC proceeded to develop a set of comparative annual energy factors that could be associated with each window, for use in any climate. A base case house was selected (see Appendix A) and the winter and summer energy performance of many windows was simulated for many climates. These seasonal energy-use numbers, which vary widely by climate, were converted into a set of energy-saving factors (one for heating and one for cooling). Each factor represents the energy-use relationship between a house with a certain type of windows compared with a base case house using clear, single-glazed, aluminum-framed windows. When the calculated energy performance was converted to these comparative factors, the variation by climate largely disappeared. This comparative seasonal performance for a particular window essentially represents the percent savings relative to the base case.

The winter savings indicator is referred to as the Heating Rating (HR) and the summer savings indicator is the Cooling Rating (CR). Figure 6-9 shows the heating and cooling factors

---

**ENERGY STAR**

Another method of informing consumers about the energy performance of windows is the Energy Star program being developed by the U.S. Department of Energy and the Environmental Protection Agency. Windows that provide superior heating and cooling performance will be labeled with the Energy Star logo (similar to programs for energy-efficient appliances). This simple consumer-friendly indicator is intended to complement the NFRC annual performance ratings.

---

Figure 6-8. Annual energy performance of four window types.

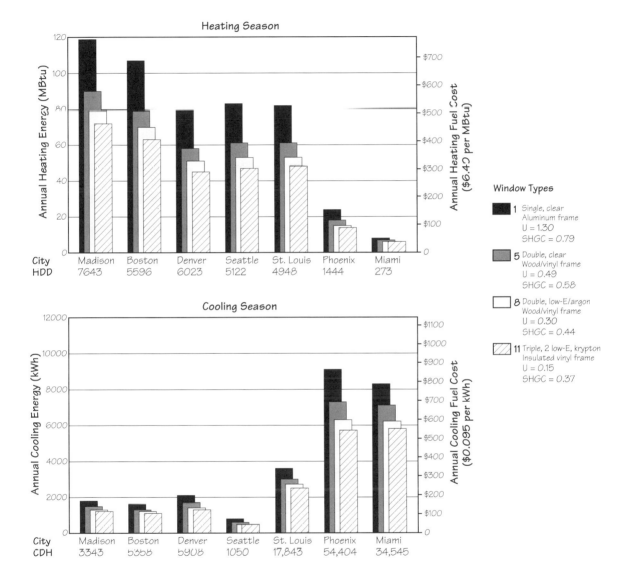

Note: The annual energy performance figures shown here are for a typical 1540 sq ft house. U-factor and SHGC are for total window including frame. House and windows are described in Appendix A. MBtu=millions of Btu, kWh=kilowatt hours. HDD=heating degree days. CDH=cooling degree hours.

Figure 6-9. Characteristics of typical window types and annual energy factors.

| Window description | Overall Window Characteristics | | | | Energy Factors* | |
| | U-factor (Btu/hr-ft²-°F) | Solar Heat Gain Coefficient | Visible Transmittance | Air Leakage (cfm/ft²) | Heat | Cool |
|---|---|---|---|---|---|---|
| 1 Single glass Aluminum frame no thermal break | 1.30 | 0.79 | 0.69 | 0.98 | 0 | 0 |
| 2 Single glass–bronze Aluminum frame no thermal break | 1.30 | 0.69 | 0.52 | 0.98 | -2 | 8 |
| 3 Double glass Aluminum frame thermal break | 0.64 | 0.65 | 0.62 | 0.56 | 19 | 12 |
| 4 Double glass, bronze Aluminum frame thermal break | 0.64 | 0.55 | 0.47 | 0.56 | 17 | 20 |
| 5 Double glass Wood or vinyl frame | 0.49 | 0.58 | 0.57 | 0.56 | 24 | 18 |
| 6 Double glass, bronze Wood or vinyl frame | 0.49 | 0.48 | 0.43 | 0.56 | 22 | 25 |
| 7 Double glass Low-E, argon Wood or vinyl frame | 0.33 | 0.55 | 0.52 | 0.15 | 32 | 19 |
| 8 Double glass Low-E, argon Wood or vinyl frame | 0.30 | 0.44 | 0.56 | 0.15 | 32 | 27 |
| 9 Double glass Selective low-E, argon Wood or vinyl frame | 0.29 | 0.32 | 0.51 | 0.15 | 30 | 36 |
| 10 Double glass Selective low-E, argon Wood or vinyl frame | 0.31 | 0.26 | 0.31 | 0.15 | 27 | 40 |
| 11 Triple glass 2 low-E, krypton Insulated vinyl frame | 0.15 | 0.37 | 0.48 | 0.08 | 38 | 33 |
| 12 Triple glass Wood or vinyl frame | 0.34 | 0.52 | 0.53 | 0.08 | 32 | 22 |

*All values in this table are for the whole window and include the effects of the frame. See Appendix A for detailed descriptions of the windows. The Energy Factors are calculated based on NFRC-900 procedure, first edition (November 2, 1995) and are being used as a basis for the NFRC Heating Rating (HR) and Cooling Ratiing (CR).

that will be used to generate HR and CR values for the twelve window systems referenced throughout the book (see Appendix A for details about these windows). Larger values represent greater energy savings and reduced operating costs. The best window technology available today would have factors as high as 45 or 50 for both heating and cooling.

The final format of the HR and CR rating system is under development. Once this process is complete, HR and CR values promise to be better indicators of relative energy use than the basic window properties such as U-factor. However, they will still be a comparative performance indicator similar to miles-per-gallon ratings for automobiles. The use of computer tools like RESFEN is required for more accurate calculation of specific annual energy use or cost savings, as explained below (Sullivan et al., 1992).

## Using a Simplified Computer Program (RESFEN)

Rating systems such as HR and CR are based on computer calculations of energy performance, but they still have limitations because many simplified assumptions are built into the calculations. With the use of computer programs, it is possible to remove most of the limitations of the HR/CR rating system and generate energy savings values for any set of windows in a specific house. In this case, the user defines the house with a series of selections from a menu: location, heating and cooling system type and efficiency, utility rates, floor area, window area, window orientation, interior/exterior shading, etc. A specific window or set of windows for each orientation is selected and specified by their U-factor, SHGC, and air leakage rate. The program then calculates the annual energy use and cost in a matter of seconds.

As with all simulation programs, there are still assumptions and approximations that must be understood, and there is a short learning curve associated with using the program. It is anticipated that the RESFEN program will be approved by NFRC for those who are willing or able to invest a little more time and effort in the window selection process. RESFEN has also been used in an electronic kiosk form at a window store. The rapidly evolving interest nationwide in delivering information electronically is making selection tools like RESFEN available to homeowners and design professionals over the World Wide Web. Check Appendix D for more information about access to RESFEN.

U.S. heating climate zone. (Boundaries of the climate zones are approximate.)

*Example 1: Window Selection in a Heating Climate*

Using a computer simulation program, four possible window choices are compared for a typical house in Madison, Wisconsin. In addition, the energy use for the same house with poor windows is shown in the first bar of Figure 6-10 to provide a comparison to an older, existing structure with single-glazed windows. Window A in Figure 6-10 is a typical clear, double-glazed unit–the most common cold-climate window installed in the U.S. during the period from 1970 to about 1985. Window B has a high-transmission low-E coating, while Window C has a spectrally selective low-E coating. Window B is designed to reduce winter heat loss (low U-factor) and provide winter solar heat gain (high SHGC). Window C also reduces winter heat loss (low U-factor) but it reduces solar heat gain as well (low SHGC). Window D, with triple glazing and two low-E coatings, is representative of the most efficient window on the market today with respect to winter heat loss (very low U-factor).

The figure illustrates that there are significant savings in annual heating costs by using windows with low U-values (Windows B and C) instead of double-glazed, clear units (Window A) or the single-glazed case. The high-transmission low-E unit (Window B) is slightly better than the spectrally selective low-E unit (Window C) in heating season performance, but Window C is clearly better during the cooling season. The triple-glazed unit (Window D), with its very low U-value, results in even greater heating season savings.

To make a window selection based on this energy performance data, it is necessary to factor in the other issues discussed throughout this chapter. Improvements in energy performance must be weighed against both initial and life-cycle costs. In addition, there are benefits of greater comfort with reduced risk of condensation. The benefit of reducing cooling costs in this climate must be examined in terms of whether air conditioning is installed in the house; however, the increased comfort of a window with a low SHGC is a factor to be considered whether or not the homeowner is paying for cooling.

In applying these typical results to your particular situation, remember that our example is a relatively small house (1500 sq ft) with an average amount of window area (231 sq ft). The fuel and electricity rates shown on the figures are national averages. Instead of drawing conclusions from average con-

## Figure 6-10: Comparing windows in a heating climate

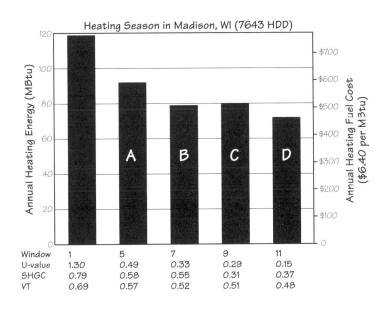

### Heating Season in Madison, WI (7643 HDD)

Annual Heating Energy (MBtu) — Annual Heating Fuel Cost ($6.40 per MBtu)

| Window | 1 | 5 | 7 | 9 | 11 |
|---|---|---|---|---|---|
| U-value | 1.30 | 0.49 | 0.33 | 0.29 | 0.15 |
| SHGC | 0.79 | 0.58 | 0.55 | 0.31 | 0.37 |
| VT | 0.69 | 0.57 | 0.52 | 0.51 | 0.48 |

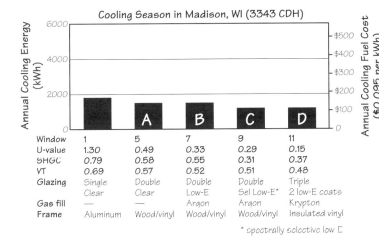

### Cooling Season in Madison, WI (3343 CDH)

Annual Cooling Energy (kWh) — Annual Cooling Fuel Cost ($0.095 per kWh)

| Window | 1 | 5 | 7 | 9 | 11 |
|---|---|---|---|---|---|
| U-value | 1.30 | 0.49 | 0.33 | 0.29 | 0.15 |
| SHGC | 0.79 | 0.58 | 0.55 | 0.31 | 0.37 |
| VT | 0.69 | 0.57 | 0.52 | 0.51 | 0.48 |
| Glazing | Single Clear | Double Clear | Double Low-E | Double Sel Low-E* | Triple 2 low-E coats |
| Gas fill | — | — | Argon | Argon | Krypton |
| Frame | Aluminum | Wood/vinyl | Wood/vinyl | Wood/vinyl | Insulated vinyl |

\* spectrally selective low E

Note: The annual energy performance figures shown here are for a typical 1540 sq ft house with 231 sq ft of window area (15% of floor area). The windows are equally distributed on all four sides of the house and are unshaded. U-factor, SHGC, and VT are for the total window including frames. House and windows are described in Appendix A. HDD=heating degree days. CDH=cooling degree hours. kWh=kilowatt hours. Mbtu=millions of Btu. The fuel and electricity prices represent national averages.

**A.** Clear double glazing

**B.** High-transmission low-E coating

**C.** Spectrally selective low-E coating

**D.** Triple glazing with low-E coatings

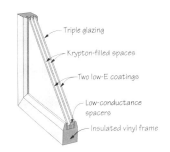

165

ditions such as these, the best way to compare different windows is by using a computer tool such as RESFEN where you can base decisions on your own house design and fuel costs for your area.

## Example 2: Window Selection in a Cooling Climate

Similar to the heating climate example above, a computer simulation program is used to compare four possible window choices for a typical house in Phoenix, Arizona. Again, the energy use for the same house with poor windows is shown in the first bar of Figure 6-11 to provide a comparison to an older, existing structure with single-glazed windows. Window A in Figure 6-11 is a typical clear, double-glazed unit. Window B, with bronze-tinted glass, represents a traditional approach to reducing solar heat gain (note the somewhat reduced SHGC accompanied by a significant reduction in daylight–lower VT). Window C represents the relatively new technology of using a spectrally selective low-E coating (a low SHGC combined with a relatively high VT). Window D, which combines a spectrally selective low-E coating with tinted glass, represents further reduction in summer heat gain (very low SHGC), but at the cost of losing daylight as well (low VT).

U.S. cooling climate zone. (Boundaries of the climate zones are approximate.)

Figure 6-11 illustrates that there are significant savings in annual cooling costs by using windows with low solar heat gain coefficients (Windows C and D) instead of double-glazed, clear units or traditional bronze-tinted glass (Windows A and B). Savings are even greater when compared to the single-glazed case which is common in many existing homes of warmer regions. Windows C and D, with their low U-values, also reduce heating costs in a warm climate where there is some heating required.

Just as with the heating-climate example, making a window selection in a cooling climate based on this energy performance data must include the other issues discussed throughout this chapter. Improvements in energy performance must be weighed against both initial and life-cycle costs. In addition, there are benefits of greater comfort in both summer and winter. The conclusion from this might be that Windows C and D are almost equal in terms of energy performance. A critical factor then becomes the amount of daylight they allow. If maximizing light and view is your goal, then Window C is the obvious choice. Window D might be selected if glare control is an overriding concern.

## Figure 6-11: Comparing windows in a cooling climate

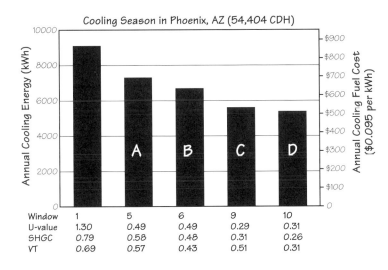

**Cooling Season in Phoenix, AZ (54,404 CDH)**

| Window | 1 | 5 | 6 | 9 | 10 |
|---|---|---|---|---|---|
| U-value | 1.30 | 0.49 | 0.49 | 0.29 | 0.31 |
| SHGC | 0.79 | 0.58 | 0.48 | 0.31 | 0.26 |
| VT | 0.69 | 0.57 | 0.43 | 0.51 | 0.31 |

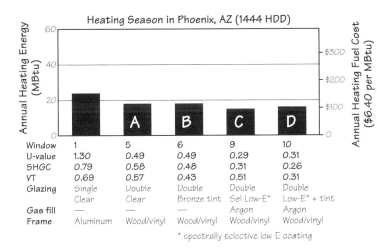

**Heating Season in Phoenix, AZ (1444 HDD)**

| Window | 1 | 5 | 6 | 9 | 10 |
|---|---|---|---|---|---|
| U-value | 1.30 | 0.49 | 0.49 | 0.29 | 0.31 |
| SHGC | 0.79 | 0.58 | 0.48 | 0.31 | 0.26 |
| VT | 0.69 | 0.57 | 0.43 | 0.51 | 0.31 |
| Glazing | Single Clear | Double Clear | Double Bronze tint | Double Sel Low-E* | Double Low-E* + tint |
| Gas fill | — | — | — | Argon | Argon |
| Frame | Aluminum | Wood/vinyl | Wood/vinyl | Wood/vinyl | Wood/vinyl |

\* spectrally selective low E coating

Note: The annual energy performance figures shown here are for a typical 1540 sq ft house with 231 sq ft of window area (15% of floor area). The windows are equally distributed on all four sides of the house and are unshaded. U-factor, SHGC, and VT are for the total window including frames. House and windows are described in Appendix A. HDD=heating degree days. CDH=cooling degree hours. kWh=kilowatt hours. Mbtu=millions of Btu. The fuel and electricity prices represent national averages.

**A.** Clear double glazing

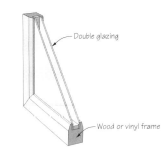

**B.** Double glazing with bronze tint

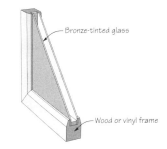

**C.** Spectrally selective low-E coating

**D.** Spectrally selective low-E coating with tinted glass

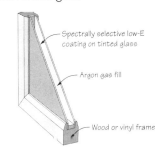

167

As noted for the heating-climate example, consider that this is for a relatively small house (1500 sq ft) with an average amount of window area (231 sq ft). The fuel and electricity rates shown on the figures are national averages.

### Example 3: Window Selection in a Mixed Climate

The previous two examples have focused on the regions of more extreme climate in the United States. In terms of analyzing energy performance, these climates are easier to address because one season clearly predominates, so decisions are clearly weighted in favor of winter heating in the north and summer cooling in the south. The great area in between these extremes is often referred to as a mixed heating and cooling climate zone. In these cases, the relative importance of the heating and cooling season performance will vary with location, utility costs, etc.

The comments for the previous heating and cooling climate examples all apply to some degree in a mixed climate. Because heating and cooling costs must be balanced in mixed climates and then combined with all of the other selection factors, it is important to use a reliable computer tool such as RESFEN where you can base decisions on your own house design and fuel costs for your area.

## Peak Heating and Cooling Loads

High-performance windows not only provide reduced annual heating and cooling bills; they reduce the peak heating and cooling loads as well. This has benefits for the homeowner, in that the size of the heating or cooling system may be reduced, and it also benefits the electrical utilities, in that load factors are reduced during the peak times in summer.

The peak load for a building is the maximum requirement for heating or cooling at one time. These loads determine the size of the furnace, heat pump, air conditioner, and fans that must be installed. The peak heating load represents how much heat must be delivered from the furnace at the coldest outdoor temperature in order to maintain a given interior temperature (typically, 70°F/21°C). Similarly, the peak cooling load represents how much air-conditioning capacity is required to maintain a given interior temperature (typically, 75°F/24°C) during the hottest summer conditions.

U.S. mixed heating and cooling climate zone. (Boundaries of the climate zones are approximate.)

Figure 6-12. Peak heating load for six window types.

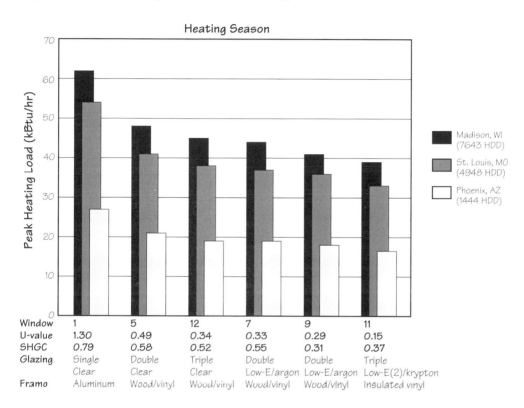

**Heating Season**

| Window | 1 | 5 | 12 | 7 | 9 | 11 |
|---|---|---|---|---|---|---|
| U-value | 1.30 | 0.49 | 0.34 | 0.33 | 0.29 | 0.15 |
| SHGC | 0.79 | 0.58 | 0.52 | 0.55 | 0.31 | 0.37 |
| Glazing | Single Clear | Double Clear | Triple Clear | Double Low-E/argon | Double Low-E/argon | Triple Low-E(2)/krypton |
| Frame | Aluminum | Wood/vinyl | Wood/vinyl | Wood/vinyl | Wood/vinyl | Insulated vinyl |

Legend:
- Madison, WI (7643 HDD)
- St. Louis, MO (4948 HDD)
- Phoenix, AZ (1444 HDD)

Note: The peak load figures shown here are for a typical 1540 sq ft house. U-factor and SHGC are for total window including frame. House and windows are described in Appendix A. HDD=heating degree days.

Figures 6-12 and 6-13 illustrate typical reductions in peak loads that occur with different window types used in a typical house. In these examples, there is a 10 to 20 percent reduction in peak heating and cooling loads by changing from clear double glazing (window 5) to windows using low-E coatings and low-conductance gases (windows 7, 9, 10, and 11). The consumer can benefit directly from peak load reduction, because heating and cooling systems can be sized smaller, resulting in initial cost savings in some cases.

The greatest economic benefits come from reductions of cooling loads in hot climates and heating loads in cold climates where heat pumps are used to provide space heat. In

Figure 6-13. Peak cooling load for six window types.

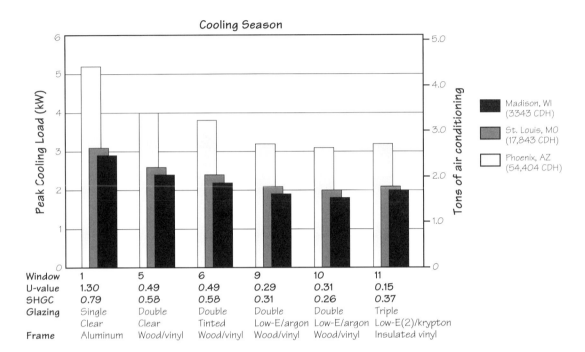

| Window | 1 | 5 | 6 | 9 | 10 | 11 |
|---|---|---|---|---|---|---|
| U-value | 1.30 | 0.49 | 0.49 | 0.29 | 0.31 | 0.15 |
| SHGC | 0.79 | 0.58 | 0.58 | 0.31 | 0.26 | 0.37 |
| Glazing | Single Clear | Double Clear | Double Tinted | Double Low-E/argon | Double Low-E/argon | Triple Low-E(2)/krypton |
| Frame | Aluminum | Wood/vinyl | Wood/vinyl | Wood/vinyl | Wood/vinyl | Insulated vinyl |

Note: The peak load figures shown here are for a typical 1540 sq ft house. U-factor and SHGC are for total window including frame. House and windows are described in Appendix A. CDH=cooling degree hours. Tons of air conditioning is calculated based on a unit with COP = 3.0.

new construction, reducing peak loads might also lead to using smaller fans and ducts. High-performance windows can be an important part of a larger package of energy efficiency measures that combine to produce large savings in space conditioning equipment.

Lowering the peak cooling load is a goal of most electric utilities. Spectrally selective insulating glazing has the ability to lower electric utility load factors during the peak summer use in warmer climates (McCluney, 1993). If peak cooling loads are minimized, additional generating capacity is not required. This directly benefits the utility company, and indirectly the consumer, by keeping rates down. Some electric utilities provide financial incentives to builders and homeowners to reduce peak loads, which can offset some of the incremental cost of higher-performance windows. Check with your local utility to see if such incentives are available.

Peak heating and cooling loads are not rated separately on the NFRC label at this time. When the annual performance ratings are on the label, higher FHR and FCR numbers will generally result in lower peak loads.

## Long-Term Maintenance of Energy Performance

Another energy-related consideration in selecting windows is the long-term ability of the window to maintain its initial energy performance. This depends upon whether the features of the window that contribute to energy savings are stable and durable or whether they may change over time, gradually reducing energy performance. The window U-factor might change if there is seal failure and loss of argon or krypton gas from the space between layers of glass. The U-factor could also be affected if there is any change or degradation in low-E coatings, frame conductivity, or suspended films. The solar heat gain coefficient (SHGC) might change if properties of coatings or tints change, or laminated or suspended films degrade. Air leakage or infiltration through windows may increase over time with any of the following conditions: wear of the weatherstrip, weatherstrip material changes, frame and sash stability (e.g., warping), frame sealant loss, and ineffective hardware performance.

It is difficult for a consumer to assess the likelihood that any of these possible conditions will occur. Except for the most obvious failure, some of them may be hard to detect without testing. The National Fenestration Council is in the process of developing ratings for long-term energy performance, which may address some of these issues. Until such ratings are available, a purchaser should ask about product warranties and talk to homeowners who have previously purchased similar units.

## The Role of Building Codes and Standards

The importance of building codes and standards cannot be overlooked in the process of ensuring the quality and performance of windows. The home designer and builder must follow the codes in order to receive approval to build. The homeowner may not be directly aware of codes and standards, but they indirectly influence the choices on the market and protect the consumer by setting minimum standards. Building codes address a number of safety and performance issues, such as using tempered glass near stairways and doors and requiring a certain penetration resistance in hurricane areas.

With respect to energy performance, building codes influence window selection in two ways. First, they may prescribe minimum levels for U-factors, solar heat gain coefficients, and air leakage. They also can prescribe a maximum window area, which is based on the premise that windows are less energy efficient than the rest of the envelope and therefore should be minimized. This somewhat out-of-date approach is usually superseded by a trade-off procedure or a performance standard that permits a particular house to meet an overall energy consumption level, but allows flexibility in the type, number, and efficiency of windows as well as other components. In effect, this performance-based approach allows the designer or homeowner to receive credit for using better-performance windows. For example, using larger areas of some low-E argon-filled windows or superwindow assemblies does not result in any significant energy penalty.

For either prescriptive or performance-based codes to work, there must be a reliable way of rating product performance. Meeting a certain U-factor, for example, cannot be based on a single manufacturer's own tests and calculations method. This is where a standardized industry rating, provided by the National Fenestration Rating Council (NFRC), is critical to ensuring code compliance as well as consumer protection. Increasingly, building codes such as the Model Energy Code for the United States and many state energy codes require an NFRC label on the product to assist the manufacturer in complying with the respective prescriptive or performance-based requirements.

## COST CONSIDERATIONS

In evaluating the cost of a window unit, the aesthetic, functional, and energy performance characteristics come together. Without the proper understanding of total window performance, costs may be evaluated based on one or two facts—the initial cost of the unit and, possibly, a characteristic such as U-value (if it is available). To make a better decision about the costs and benefits of a particular window, it is useful to think of the life-cycle cost of the unit. There are both monetary and nonmonetary costs and benefits. Even if the decision is made to base window selection on monetary costs and benefits only, the following factors should be included:

1. Initial Cost

   This is the most obvious of the factors and is included in virtually any analysis. For new construction, initial cost may be defined as only the cost of the window unit itself, since the installation cost will generally be equal for any similar window. In a retrofit situation, the comparison between an existing window and a replacement window usually should include the installation cost as part of the project cost analysis. If the windows must be replaced because of failure of the existing units, then the comparative cost analysis might only include the relative costs of the alternative new window units, not the installation. The cost of a specific window from a particular manufacturer may vary considerably between alternative suppliers. Since window purchases are often large investments, it pays to shop aggressively for the best price and take advantage of any sales or discounts.

2. Cost of Exterior and Interior Window Treatments

   At first glance, window treatments do not seem to be an integral part of comparing the cost of window units. With newer high-performance windows, however, they may be quite relevant to consider. Many of the higher-performance windows (using low-E coatings, gas fills, and low-conductance frames) do the work of various shading and insulating components that were necessary to achieve comfort in the past. If choosing a better window reduces the need for some external shading devices or interior blinds or drapes (or allows the purchase of a simpler, cheaper version), then this is part of the total cost/benefit picture. Of course, elements such as blinds or drapes may have an additional function of providing privacy or glare control, but this is not always the case.

3. Cost of Maintenance

   In the past, window maintenance has represented a significant cost and effort. Materials such as vinyl and aluminum, as well as claddings and more durable finishes on wood windows, have greatly reduced exterior maintenance. When choosing between products that

---

**Cost**
- Initial cost of window units and installation
- Cost of interior and exterior window treatments
- Cost of maintenance
- Frequency of replacement
- Resale value
- Initial cost of heating and cooling system
- Annual cost of heating and cooling energy

have different maintenance requirements (i.e., continued painting of an existing window versus a new, maintenance-free window), consider the cost and frequency of this activity over the life of the window.

4. Frequency of Replacement

Some windows simply last longer than others because of their quality and durability. The life of a window unit is related to its design, materials, and workmanship. In assessing life cycle costs for a window purchase, the frequency of replacement should be included as well.

5. Resale Value

It is well known that special features such as a large bay window, skylights, or a glazed sun porch can add value to a home. High-performance windows may also add to the resale value of a house. Energy-efficient windows can be attractive selling features, and lead to higher appraisals. Homes with lower annual energy costs provide some protection from future fuel cost increases.

6. Cost of Heating and Cooling System

In some cases, choosing a higher-performance window reduces the peak heating and cooling loads sufficiently so that the mechanical system size can be smaller, resulting in initial cost savings on furnaces and air-conditioning units. Utilities may offer rebates as incentives to purchase windows which result in lower cooling loads and thus lower peak electric demand. (See previous section, Peak Heating and Cooling Loads.)

7. Annual Cost of Heating and Cooling Energy

This is a very important but often elusive cost to determine. Earlier in this chapter methods for determining the impact of window choices on annual energy costs were described. Figures 6-14 through 6-16 show examples of the heating and cooling costs for a typical 1540-square-foot house in three climates. What is important to note are the relative comparisons between the different climates, glazings, and seasons. These dollar amounts are only relevant for the specific house and the fuel cost assumptions used in the calculation. Once energy cost savings are determined for a particular win-

Figure 6-14. Annual energy cost comparison for a typical house using several window types in Madison, Wisconsin.

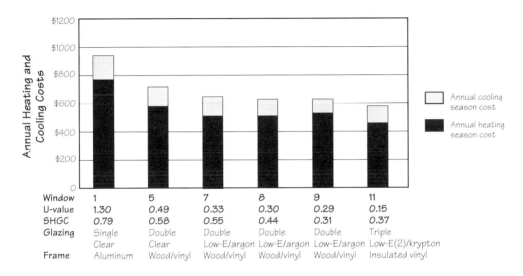

| Window | 1 | 5 | 7 | 8 | 9 | 11 |
|---|---|---|---|---|---|---|
| U-value | 1.30 | 0.49 | 0.33 | 0.30 | 0.29 | 0.15 |
| SHGC | 0.79 | 0.58 | 0.55 | 0.44 | 0.31 | 0.37 |
| Glazing | Single Clear | Double Clear | Double Low-E/argon | Double Low-E/argon | Double Low-E/argon | Triple Low-E(2)/krypton |
| Frame | Aluminum | Wood/vinyl | Wood/vinyl | Wood/vinyl | Wood/vinyl | Insulated vinyl |

Figure 6-15. Annual energy cost comparison for a typical house using several window types in St. Louis, Missouri.

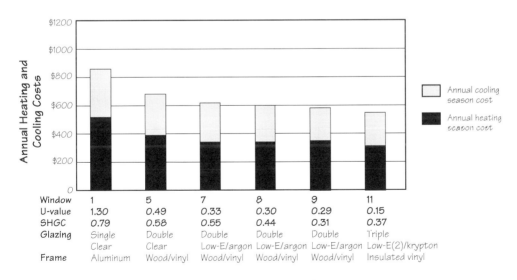

| Window | 1 | 5 | 7 | 8 | 9 | 11 |
|---|---|---|---|---|---|---|
| U-value | 1.30 | 0.49 | 0.33 | 0.30 | 0.29 | 0.15 |
| SHGC | 0.79 | 0.58 | 0.55 | 0.44 | 0.31 | 0.37 |
| Glazing | Single Clear | Double Clear | Double Low-E/argon | Double Low-E/argon | Double Low-E/argon | Triple Low-E(2)/krypton |
| Frame | Aluminum | Wood/vinyl | Wood/vinyl | Wood/vinyl | Wood/vinyl | Insulated vinyl |

Note: The cost figures shown here are based on annual energy performance calculations for a typical 1540 sq ft house. Heating costs are based on a national average cost of $6.40 per MBtu for natural gas. Cooling costs are based on a national average cost of $0.095 per kWh for electricity. U-factor and SHGC are for total window including frame. House and windows are described in Appendix A.

Figure 6-16. Annual energy cost comparison for a typical house using several window types in Phoenix, Arizona.

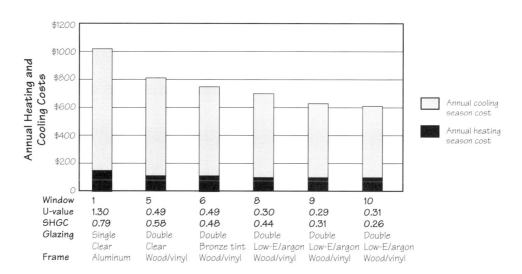

| Window | 1 | 5 | 6 | 8 | 9 | 10 |
|---|---|---|---|---|---|---|
| U-value | 1.30 | 0.49 | 0.49 | 0.30 | 0.29 | 0.31 |
| SHGC | 0.79 | 0.58 | 0.48 | 0.44 | 0.31 | 0.26 |
| Glazing | Single Clear | Double Clear | Double Bronze tint | Double Low-E/argon | Double Low-E/argon | Double Low-E/argon |
| Frame | Aluminum | Wood/vinyl | Wood/vinyl | Wood/vinyl | Wood/vinyl | Wood/vinyl |

Note: The cost figures shown here are based on annual energy performance calculations for a typical 1540 sq ft house. Heating costs are based on a national average cost of $6.40 per MBtu for natural gas. Cooling costs are based on a national average cost of $0.095 per kWh for electricity. U-factor and SHGC are for total window including frame. House and windows are described in Appendix A.

dow type, it must be remembered that this is an annually recurring savings.

Figure 6-14 illustrates simple patterns that would be expected in both the heating and cooling seasons. As the windows are improved, annual energy costs are reduced. Also, the relative importance of heating and cooling is obviously linked to climate. Combining the heating and cooling energy costs reveals that total annual costs appear similar in all climates. The analysis in Figure 6-14 is based on national average costs for natural gas and electricity. You can make an approximate adjustment to these values to reflect your local energy costs. For example, if your electricity costs $0.12 per kWh, multiply the cooling season costs in Figure 6-14 by 0.12/0.095 = 1.26 to obtain your annual costs. A computer program such as RESFEN is designed to do this type of analysis quickly and easily for many major cities.

One interesting aspect of this analysis is the difference between certain low-E and spectrally selective coatings. Windows 8 and 9 in the figures have similar U-factors but window 9 has a much lower SHGC than window 8. Window 9 is clearly the best performer in a cooling-dominated climate like Phoenix; however, windows 8 and 9 have similar combined annual costs in St. Louis and Madison, although the different SHGC reflect different seasonal patterns. The lower SHGC of window 9 reduces cooling costs but increases heating costs, because less passive solar gain is available in winter.

## Financing Energy-Efficient Windows

Windows that are purchased in new homes or as part of a major renovation are typically paid for through a mortgage, home-equity loan, or some other form of long-term financing that distributes the purchase and installation cost over many monthly payments. The monthly energy savings from high-performance windows can offset or even exceed these payments. Here is a simple example:

A homeowner chooses an energy-efficient window upgrade option (low-E, gas-fill glazing) in a new house, at an additional cost of $500 (about $2 per square foot) above the standard cost of windows. This glazing provides energy savings of about $100 per year, or an average of $8.33 per month (see Figure 6-14: savings in Madison for window 4 versus window 3). The additional purchase cost of $500 is amortized as part of an 8 percent loan over 30 years. The added monthly payment, including interest, is $3.67, which is less than the average monthly energy savings, thus providing a net "profit" on this efficiency investment of $4.66 per month. More important, from the first day the house is occupied and thoughout the life of the mortgage, the owner has lower monthly living costs (mortgage plus utilities), as well as the advantage of a more comfortable home.

## SUMMARY

Our lives are filled with many purchasing decisions. Some are tightly constrained by economic factors, while others may be based more on appearance, style, and comfort. The appropriate balance depends on the unique priorities and circumstances of each individual.

Figure 6-17 summarizes all the selection considerations discussed in this chapter. It indicates the source of information needed and the basis for making a choice. It is notable that the NFRC label is a useful source of information for many considerations. At this time, NFRC labels are only available in some states, but the number is expected to grow rapidly.

Once a life-cycle analysis of monetary benefits has been completed, it is essential to recognize that this is only part of the picture. Subjective issues of appearance, comfort, and performance are all part of the overall cost/benefit analysis, and homeowners must place their own values on these considerations just as they do when making other purchases.

Figure 6-17. Summary checklist of window selection considerations.

| Considerations | Information Source | Basis for Selection |
| --- | --- | --- |
| **Appearance** | | |
| • Style | Product literature/store visit | Appropriate design, personal preference |
| • Frame materials | Product literature/store visit | Appropriate design, personal preference |
| • Size and shape | Product literature/store visit | Appropriate design, personal preference |
| • Glass color and clarity | Store visit | Visual inspection |
| **Function** | | |
| • Daylight | NFRC label (VT) | Higher VT is better for daylight |
| • Glare control | NFRC label (VT) | Choose lower VT or control with design |
| • Fading | Product literature | Lower UV transmittance is better |
| • Thermal comfort | NFRC label (U-factor) | Lower U-factor is better (below 0.4) |
| • Resistance to condensation | NFRC label (U-factor) | Lower U-factor is better (below 0.4) |
| • Ventilation/operating type | Product literature/store visit | Hinged have more open area than sliders |
| • Sound control | Product literature (STB or OITC) | Lower values are better |
| • Maintenance | Product literature (ext. frame) | Exposed wood requires more than others |
| • Durability (warranty) | Product literature/store visit | Warranty and experience of others |
| **Energy performance** | | |
| • Basic thermal properties | | |
|     U-factor | NFRC label (U-factor) | Lower U-factor is better (below 0.4) |
|     SHGC | NFRC label (SHGC) | Lower SHGC for cooling, higher SHGC for solar gain |
|     Air leakage | NFRC label (AL–future) | Below 0.56 cfm/sq ft (lower is better) |
| • Annual heating and cooling season performance | NFRC label (HR & CR–future) | Higher HR and CR are better |
| • Peak load impacts | NFRC label (future?) | Higher HR and CR = lower peak loads |
| • Long-term ability to maintain energy performance | NFRC label (future?) | Warranty and experience of others |
| **Cost** | | |
| • Initial cost of window units | Contractor or store | Lowest cost with desired performance |
| • Cost of interior and exterior window treatments | Contractor or store | Include in cost analysis if required to obtain adequate energy performance |
| • Cost of maintenance | Estimate | Include in life-cycle cost analysis |
| • Frequency of replacement | Estimate | Include in life-cycle cost analysis |
| • Resale value | Estimate | Include in life-cycle cost analysis |
| • Initial cost of heating and cooling system | Contractor | Include in life-cycle cost analysis |
| • Annual cost of heating and cooling energy | NFRC label (HR and CR) or computer simulation (RESFEN) | Include in life-cycle cost analysis |
| • Life-cycle cost | Calculate using factors above | Lowest life-cycle cost |

# APPENDIX A

# Energy Performance Calculation Assumptions

In order to understand the impact of window selection on annual energy use, researchers performed thousands of computer simulations of houses with a wide range of characteristics in several U.S. climates. The simulation program used was DOE 2.1E, which is considered one of the standards for building simulation in the United States. The results of these simulations appear throughout the book. The basic assumptions concerning the house and window units are described below.

### The Prototypical House

In order to conduct a series of computer simulations where various conditions are changed to determine their impact, it is necessary to begin with a prototypical house that serves as the base case. Whether or not this house matches a particular design is not important at this point. The main purpose is to discover the relative impact of changing window types and conditions. The characteristics of the prototypical house used in the simulations are shown in Figure A-1.

### Window Units

Twelve window units were choscn for the simulations. They reflect a broad range of U-factor and solar heat gain coefficients (SHGC). These twelve window types do not include every unit available on the market but the U-factor and SHGC for most units is relatively close to one of these. Detailed descriptions of the window characteristics used in the simulations appear in Figure A-2.

Figure A-1. Characteristics of the prototypical house used in computer simulations.

## Building

| | |
|---|---|
| Floor area | 1540 square feet |
| Insulation levels | R-30 ceilings |
| | R-19 walls |
| | R-19 floor |
| Foundation type | Crawl space |
| House infiltration | 0.5 air changes per hour |
| Natural ventilation | 10 air changes per hour |
| | enthalpic venting |
| Thermal mass in building | 3.5 lbs per sq ft for structure |
| | 8.0 lbs per sq ft for furnishings |

## Windows

| | |
|---|---|
| Fenestration area | 231 square feet |
| | (15% of floor area) |
| Window orientation | Equal (57.75 sq ft on each side) |
| External shading | None |
| Internal shades/blinds | None |

## Mechanical System

| | |
|---|---|
| Mechanical system type | Gas furnace with central |
| | air conditioning |
| HVAC efficiency | Heating system efficiency = 78% |
| | Air conditioning system: 10 SEER |
| Thermal zones | One |
| Internal loads | 54 kBtu per day |
| Thermostat settings | Heating = 70°F,  Cooling = 78°F |
| Seasonal plant cutout | None |

Figure A-2. Characteristics of window types used in simulations.

| Characteristic | Window | | | | | |
| --- | --- | --- | --- | --- | --- | --- |
| | 1 | 2 | 3 | 4 | 5 | 6 |
| General glazing description | Single-glazed clear | Single-glazed bronze | Double-glazed clear | Double-glazed bronze | Double-glazed clear | Double-glazed bronze |
| Layers of glazing and spaces (outside to inside) | 1/8" clear | 1/8" bronze | 1/8" clear | 1/8" bronze | 1/8" clear | 1/8" bronze |
| | | | 1/2" air | 1/2" air | 1/2" air | 1/2" air |
| | | 1/8" clear | 1/8" clear | 1/8" clear | 1/8" clear | 1/8" clear |
| **Center-of-glass** | | | | | | |
| U-factor | 1.11 | 1.11 | 0.49 | 0.49 | 0.49 | 0.49 |
| SHGC | 0.86 | 0.73 | 0.76 | 0.62 | 0.76 | 0.62 |
| SC | 1.00 | 0.84 | 0.89 | 0.72 | 0.89 | 0.72 |
| VT | 0.90 | 0.68 | 0.81 | 0.61 | 0.81 | 0.61 |
| **Frame** | | | | | | |
| Type | Aluminum no th. break | Aluminum no th. break | Aluminum th. break | Aluminum th. break | Wood or vinyl | Wood or vinyl |
| U-factor | 1.90 | 1.90 | 1.00 | 1.00 | 0.40 | 0.40 |
| Spacer | — | — | Aluminum | Aluminum | Aluminum | Aluminum |
| **Total window** | | | | | | |
| U-factor | 1,30 | 1.30 | 0.64 | 0.64 | 0.49 | 0.49 |
| SHGC | 0.79 | 0.69 | 0.65 | 0.55 | 0.58 | 0.48 |
| VT | 0.69 | 0.52 | 0.62 | 0.47 | 0.57 | 0.43 |
| **Air leakage** | | | | | | |
| cfm/lf | 0.65 | 0.65 | 0.37 | 0.37 | 0.37 | 0.37 |
| cfm/sq ft | 0.98 | 0.98 | 0.56 | 0.56 | 0.56 | 0.56 |

Units for all U-factors are Btu/hr-ft$^2$-°F. All values for total window are based on a 2-foot by 4-foot casement window.

| Characteristic | Window | | | | | |
|---|---|---|---|---|---|---|
| | 7 | 8 | 9 | 10 | 11 | 12 |
| General glazing description | Double-glazed low-E | Double-glazed low-E | Double-glazed spectrally selective | Double-glazed spectrally selective | Triple-glazed low-E superwindow | Triple-glazed clear |
| Layers of glazing and spaces (outside to inside) | 1/8" clear | 1/8" low-E (0.08) | 1/8" low-E (0.04) | 1/8" low-E (0.10) | 1/8" low-E (0.08) | 1/8" clear |
| | 1/2" argon | 1/2" argon | 1/2" argon | 1/2" argon | 1/2" krypton | 1/2" air |
| | 1/8" low-E (0.20) | 1/8" clear | 1/8" clear | 1/8" clear | 1/8" clear | 1/8" clear |
| | | | | | 1/2" krypton | 1/2" air |
| | | | | | 1/8" low-E (0.08) | 1/8" clear |
| **Center-of-glass** | | | | | | |
| U-factor | 0.30 | 0.26 | 0.24 | 0.27 | 0.11 | 0.31 |
| SHGC | 0.74 | 0.58 | 0.41 | 0.32 | 0.49 | 0.69 |
| SC | 0.86 | 0.68 | 0.47 | 0.38 | 0.57 | 0.81 |
| VT | 0.74 | 0.78 | 0.72 | 0.44 | 0.68 | 0.75 |
| **Frame** | | | | | | |
| Type | Wood or vinyl | Wood or vinyl | Wood or vinyl | Wood or vinyl | Insulated vinyl | Wood or vinyl |
| U-factor | 0.30 | 0.30 | 0.30 | 0.30 | 0.20 | 0.30 |
| Spacer | Stainless | Stainless | Stainless | Stainless | Insulated | Stainless |
| **Total window** | | | | | | |
| U-factor | 0.33 | 0.30 | 0.29 | 0.31 | 0.15 | 0.34 |
| SHGC | 0.55 | 0.44 | 0.31 | 0.26 | 0.37 | 0.52 |
| VT | 0.52 | 0.56 | 0.51 | 0.31 | 0.48 | 0.53 |
| **Air leakage** | | | | | | |
| cfm/lf | 0.10 | 0.10 | 0.10 | 0.10 | 0.05 | 0.10 |
| cfm/sq ft | 0.15 | 0.15 | 0.15 | 0.15 | 0.08 | 0.15 |

## Summary of Annual Energy Use

Using the base case house (Figure A-1) and the twelve window units (Figure A-2), the heating and cooling energy use was calculated. The results appear in Figures A-3 and A-4.

Figure A-3. Annual heating loads for prototypical house (MBtu).

| Window | Madison 7643 HDD | Boston 5596 HDD | Denver 6023 HDD | Seattle 5122 HDD | St. Louis 4948 HDD | Phoenix 1444 HDD | Miami 273 HDD |
|---|---|---|---|---|---|---|---|
| 1 | 119.6 | 106.9 | 79.9 | 83.3 | 81.5 | 23.6 | 10.0 |
| 2 | 121.6 | 109.2 | 82.6 | 84.9 | 83.4 | 23.9 | 10.1 |
| 3 | 95.2 | 84.0 | 61.5 | 65.0 | 64.4 | 18.8 | 9.2 |
| 4 | 97.6 | 86.4 | 64.4 | 66.7 | 66.4 | 19.0 | 9.2 |
| 5 | 89.8 | 79.1 | 57.7 | 60.7 | 60.7 | 17.5 | 8.9 |
| 6 | 92.1 | 81.4 | 60.5 | 62.4 | 62.6 | 17.7 | 9.0 |
| 7 | 78.7 | 69.3 | 49.6 | 52.6 | 52.9 | 15.5 | 8.6 |
| 8 | 78.8 | 69.7 | 50.6 | 52.6 | 53.2 | 15.1 | 8.5 |
| 9 | 80.7 | 71.4 | 53.4 | 53.9 | 54.9 | 15.2 | 8.6 |
| 10 | 83.1 | 73.4 | 55.8 | 55.6 | 56.9 | 15.8 | 8.6 |
| 11 | 71.6 | 63.5 | 45.4 | 47.1 | 48.3 | 13.5 | 8.3 |
| 12 | 81.0 | — | — | — | 54.6 | 15.7 | — |
| No windows | 74.7 | 66.4 | 52.5 | 48.9 | 51.5 | 14.8 | 8.3 |

Figure A-4. Annual cooling loads for prototypical house (kWh).

| Window | Madison 3343 CDH | Boston 5358 CDH | Denver 5908 CDH | Seattle 1050 CDH | St. Louis 17,843 CDH | Phoenix 54,404 CDH | Miami 34,545 CDH |
|---|---|---|---|---|---|---|---|
| 1 | 1796 | 1583 | 2127 | 750 | 3595 | 9092 | 8276 |
| 2 | 1630 | 1436 | 1902 | 665 | 3329 | 8526 | 7652 |
| 3 | 1598 | 1409 | 1837 | 663 | 3231 | 7840 | 7492 |
| 4 | 1427 | 1261 | 1611 | 580 | 2948 | 7245 | 6842 |
| 5 | 1498 | 1326 | 1685 | 609 | 3037 | 7288 | 7091 |
| 6 | 1344 | 1193 | 1471 | 545 | 2772 | 6739 | 6468 |
| 7 | 1470 | 1275 | 1583 | 589 | 2965 | 6941 | 6815 |
| 8 | 1317 | 1150 | 1386 | 526 | 2688 | 6303 | 6220 |
| 9 | 1147 | 1024 | 1188 | 466 | 2373 | 5627 | 5551 |
| 10 | 1073 | 978 | 1121 | 438 | 2254 | 5402 | 5320 |
| 11 | 1230 | 1071 | 1247 | 492 | 2505 | 5748 | 5819 |
| 12 | 1397 | — | — | — | 2848 | 6719 | — |
| No windows | 738 | 706 | 681 | 321 | 1719 | 3730 | 3886 |

Note:  The annual energy performance figures shown here are for a typical 1540 sq ft house described in Figure A-1. Windows are described in Figure A-2. MBtu=millions of Btu, kWh=kilowatt hours. HDD=heating degree days. CDH=cooling degree hours.

# APPENDIX B

# Overview of the NFRC Program

The National Fenestration Rating Council (NFRC) was formed in 1989 to respond to a need for fair, accurate, and credible ratings for fenestration products. Fenestration products include all types of windows, skylights, and doors, both glazed and opaque: vertical sliders, horizontal sliders, casements, projecting (awning), fixed (includes nonstandard shapes), single door with frame, double and multiple doors with frames, glazed wall systems (site-built windows), skylights and sloped glazings, greenhouse/garden, dual action, and pivoted windows.

National Fenestration Rating Council

NFRC has adopted rating procedures for U-factor (NFRC 100, including 100B for doors), solar heat gain coefficient (NFRC 200), optical properties including visible light transmittance (NFRC 300), emittances (NFRC 301), and air leakage (NFRC 400). To provide certified ratings, manufacturers follow the requirements in the NFRC Product Certification Program (PCP) which involves working with laboratories accredited to the NFRC Laboratory Accreditation Program (LAP) and Independent Certification and Inspection Agencies accredited through the NFRC Certification Agency Program (CAP). The complete NFRC program, with various checks to maintain a high degree of confidence and integrity, is summarized below.

NFRC 100 was the first rating procedure approved and thus the first NFRC procedure adopted into state energy codes. Because it is also the most widely adopted, it is a good choice as an example to demonstrate the process. NFRC 100 requires the use of a combination of state-of-the-art computer simulations and improved thermal testing to determine U-factors for the whole product. Manufacturers seeking to acquire energy performance ratings for their products contact NFRC-accredited simulation laboratories. These simulation laboratories use advanced computer tools to calculate product performance ratings in accordance with NFRC 100. To become accredited, each laboratory must demonstrate competence in the use of the computer programs used in the rating system and must meet strict independence criteria. Following computer simulation, the products with the highest and lowest simulated U-factors within each product line undergo thermal testing to validate the computer simulations. The testing is performed by an NFRC-accredited testing laboratory. These laboratories have demonstrated their ability to conduct NFRC thermal tests, are periodically inspected and evaluated by the NFRC for continued competence, and are independent of any product manufacturer that they serve. Generally, if the test results are within 10 percent of the simulated values, then the simulated U-factors are considered validated, and manufacturers have product ratings for that entire product line.

The next step is product certification. NFRC has a series of checks and balances to ensure that the rating system is accurately and uniformly employed. Products and their ratings are authorized for certification by an NFRC-licensed independent certification and inspection agency (IA). The IA reviews all simulation and test information, conducts in-plant inspections, and provides secondary oversight to the manufacturer's in-house quality control program. This helps to ensure that the rated products reaching the marketplace are built in the same manner as the product samples simulated and tested, that the appropriate product ratings and labels are put on the correct products, and that the manufacturer maintains an in-house quality assurance program. Licensed IAs must demonstrate their ability to perform these services and meet strict independence criteria.

The authorization for certification means that the manufacturer is able to have the product listed in the NFRC Product Directory. However, this does not mean that a product is certified. The actual act of certification occurs when the manufacturer labels the product. Two labels are required: the temporary label (shown in Chapter 1), which contains the product ratings, and a permanent label, which allows tracking back to the IA and information in the NFRC Product Directory. In addition to informing the buyer, the temporary label provides building inspectors with the information necessary to verify energy code compliance. The permanent label provides access to energy rating information for a future owner, property manager, building inspector, lending agency, or building energy rating organization.

This process has a number of noteworthy features that make it superior to previous fenestration energy rating systems and correct past problems.

- The procedures provide a means for manufacturers to take credit for all the nuances and refinement to their product design, and a common basis for others to compare product claims.

- The involvement of independent laboratories and the IA provides architects, engineers, designers, contractors, consumers, building officials, and utility representatives with greater confidence that the information is unbiased.

- By requiring simulation and testing, there is an automatic check on accuracy. This also remedies a shortcoming of previous state energy requirements that relied on testing alone, which allowed manufacturers to perform several tests and then use the best one for code purposes.

- The certification process indicates that the manufacturer is consistently producing the product that was rated. This also corrects a past problem—that manufacturers were able to make an exceptionally high-quality sample and obtain a good rating in a test, but not consistently produce that product.

- There is now a readily visible temporary label that can be used by the building inspector to quickly verify compliance with the energy code.

- The permanent label provides future access to energy rating information.

While the program is similar for other fenestration characteristics, there are differences worth pointing out. The solar heat gain coefficient ratings (NFRC 200), which have been referenced in several codes, are based on simulation alone. The visible light transmittance and other optical properties (NFRC 300) and emittance (NFRC 301) are based on measurements by the manufacturer, with independent verification. The air leakage ratings (NFRC 400) are based on testing alone.

The NFRC publishes a directory of products eligible to be certified. For further information and to obtain copies of program documents and publications, contact:

National Fenestration Rating Council
1300 Spring Street, Suite 120
Silver Spring, MD 20910

Phone: 301-589-6372
Fax: 301-588-0854
http//:www.nfrc.org

# APPENDIX C

# Specifying Product Performance

The National Fenestration Rating Council (NFRC) was established in 1989 to develop a fair, accurate, and credible rating system for fenestration products. This was in response to the technological advances and increasing complexity of these products, which manufacturers wanted to take credit for but which cannot be easily visually verified. NFRC procedures started to be incorporated in State Energy Codes in 1992. The 1992 National Energy Policy Act provided for the development of a national rating system. The U.S. Department of Energy has selected the NFRC program and certified it as the national rating system. In addition, the NFRC procedures are now referenced in and being incorporated into the Model Energy Code and ASHRAE Standards 90.1 and 90.2. Contractors and homeowners can and should look for the NFRC label when selecting fenestration products as a way of knowing that the ratings have been developed in a nationally recognized acceptable manner with independent oversight and that the manufacturer is willing to certify the product. Designers should specify the NFRC rating procedures and certification program to ensure that their design is implemented as desired and that bidders are all competing on a common basis. Specifications are provided on the facing page as an aid to designers.

# SAMPLE SPECIFICATION FOR FENESTRATION PRODUCTS

1. U-factor for all fenestration products (windows, doors, skylights) shall include the effects of glass, sash, and frame and shall be determined in accordance with National Fenestration Rating Council (NFRC) 100. The product shall be labeled for U-factor and certified by the manufacturer in accordance with the NFRC Product Certification Program.

2. Solar Heat Gain Coefficient (SHGC) for all fenestration products (windows, doors, skylights) shall include the effects of glass, sash, and frame and shall be determined in accordance with National Fenestration Rating Council (NFRC) 200. The product shall be labeled for SHGC and certified by the manufacturer in accordance with the NFRC Product Certification Program.

3. Visible Light Transmittance for all fenestration products (windows, doors, skylights) shall include the effects of glass, sash, and frame and shall be determined in accordance with National Fenestration Rating Council (NFRC) 300. The product shall be labeled for Visible Light Transmittance and certified by the manufacturer in accordance with the NFRC Product Certification Program.

4. Air Leakage for all fenestration products (windows, doors, skylights) shall include the effects of glass, sash, and frame and shall be determined in accordance with National Fenestration Rating Council (NFRC) 400. The product shall be labeled for Air Leakage and certified by the manufacturer in accordance with the NFRC Product Certification Program.

# APPENDIX D

# Resources

This appendix includes a variety of resources that are referred to in the book as well as some additional sources of information related to windows. There are lists of organizations followed by lists of printed material directly relevant to window installation and technical standards for windows.

In seeking information concerning windows and energy-efficiency in general, there are several local resources worth investigating:

- Local utilities
- State or municipal energy agencies
- Regional Universities which may have architecture, construction, or extension programs
- Bookstores
- Publications in window sales areas
- Local chapters of the American Institute of Architects
- Local builder's associations

## Computer Simulation Programs

### RESFEN

A program for calculating the annual heating and cooling energy use and costs due to fenestration systems. RESFEN also calculates their contribution to peak heating and cooling loads. Available from:

National Fenestration Rating Council (NFRC)
1300 Spring Street, Suite 120
Silver Spring, MD USA 20910
Phone: (301) 589-NFRC

RESFEN can be downloaded at:
http://eande.lbl.gov/BTP/BTP.html
In the future, an interactive version of RESFEN will be directly accessible at this web site.

### Window

A program that calculates thermal performance of fenestration products. The heat transfer method is consistent with the rating procedure developed by the National Fenestration Rating Council. Available from:

Ruth Haynes
Lawrence Berkeley National Laboratory
Mail Stop 90-3111
1 Cyclotron Road
Berkeley, CA 94720
Fax: (510) 486-4089

### DOE-2.1E

An hourly, whole-building energy analysis program for calculating energy performance and life-cycle cost of operation. Commercial PC versions are available. Request a copy of the latest *User News* from Kathy Ellington for an up-to-date list.
Fax: (510) 486-4089

## American Professional and Manufacturer's Organizations

American Architectural Manufacturers Association (AAMA)
1827 Walden Office Square, Suite 104
Schaumberg, IL 60173
Phone: (847) 303-5664/Fax: (847) 303-5774
http://www.aamanet.org

American Institute of Architects (AIA)
1735 New York Ave.. N. W.
Washington, DC 20006
Phone: (202) 626-7300
http://www.aia.org

American National Standards Institute (ANSI)
11 West 42nd Street, 13th Floor
New York, NY 10036
Phone: (212) 642-4900/Fax: (212) 398-0023
http://www.ansi.org

American Society of Heating, Refrigerating and Air
Conditioning Engineers, Inc. (ASHRAE)
1791 Tullie Circle, NE
Atlanta, GA 30329-2305
Phone: (404) 6636-8400/Fax: (404) 321-5478
http://www.ashrae.org

American Society for Testing and Materials (ASTM)
1916 Race Street
Philadelphia, PA 19103-1187
Phone: (215) 299-5400/Fax: (215) 977-9679
http://www.astm.org

American Solar Energy Society
2400 Central Avenue, G-1
Boulder, CO 80301-2843
Phone: (303) 443-3130/Fax: (303) 443-3212
http://www.ases.org/solar

National Association of Home Builders (NAHB)
15th and M Street
Washington, DC
Phone: (202) 822-0200
http://www.nahb.com

National Fenestration Rating Council (NFRC)
1300 Spring Street, Suite 120
Silver Spring, MD USA 20910
Phone: (301) 589-NFRC
http://www.nfrc.org

National Wood Window and Door Association
1400 East Touhy Avenue, Suite G-54
Des Plaines, IL 60018
Phone: (847) 299-5200

Passive Solar Industries Council
1511 K Street, NW, Suite 600
Washington, DC 20005
Phone: (202) 628-7400/Fax: (202) 393-5043
http://www.psic.org

Primary Glass Manufacturers Council (PGMC)
3310 Harrison
Topeka, KS 66611
Phone: (913) 266-3666/Fax: (913) 266-0272
e-mail: Jimpgmc@aol.com

## Canadian and International Organizations

Canadian Home Builders' Association
200 Elgin Street, Suite 502
Ottawa, Ontario K2P 1L5
Phone: (613) 230-3060

Canadian Standards Association
178 Rexdale Boulevard
Rexdale, Ontario M9W 1R3
Phone: (416) 747-4000

Canadian Window and Door Manufacturers
Association
27 Goulbourn Avenue
Ottawa, Ontario K1N 8C7
Phone: (613) 233-9804

International Energy Agency
Solar Heating and Cooling Programme
1808 Corcoran Street, NW
Washington, DC 20009
Phone: (202) 483-2393/Fax: (202) 265-2248
http://www.arch.vuw.ac.nz/iea/index.html

International Energy Agency
Center for the Analysis and Dissemination of
Demonstrated Energy Technologies (CADDET)
http://www.ornl.gov/CADDET/caddet.html

International Standards Organization
1, rue de Varembe'
Case postale 56
CH-1211 Gene've 20
Switzerland
Phone: + 41 22 749 01 11/Fax: + 41 22 733 34 30
http://www.iso.ch/welcome.html

National Research Council of Canada
Institute for Research in Construction (IRC)
http://www.cisti.nrc.ca:80/irc/irccontents.html

Standards Council of Canada
45 O'Connor Street, Suite 1200
Ottawa, Ontario K1P 6N7
Phone: (613) 238-3222/Fax: (613) 995-4564
http://www.scc.ca

## Government, Research, and Educational Organizations

Florida Solar Energy Center (FSEC)
1679 Clearlake Road
Cocoa, FL 32922
Phone: (407) 638-1000/Fax: (407) 638-1010
http://www.fsec.ucf.edu

Lawrence Berkeley National Laboratory (LBNL)
Building Technologies Program
Energy and Environment Division
Lawrence Berkeley Laboratory, Berkeley, CA 94721
Phone: (510) 486-6845/Fax: (510) 486-4089
http://eande.lbl.gov/BTP/BTP.html

Minnesota Building Research Center (MNBRC)
University of Minnesota
1425 University Avenue SE
Minneapolis, MN 55455
Phone: 612-626-7819/Fax: 612-626-7242
http://www.umn.edu/mnbrc

National Renewable Energy Laboratory
Center for Buildings and Thermal Energy Systems
1617 Cole Blvd.
Golden, CO 80401
(303) 384-7520
Fax: (303) 384-7540
http://nrel.gov

National Technical Information Service (NTIS)
http://www.fedworld.gov/ntis/ntishome.html

Oak Ridge National Laboratory (ORNL)
Building Envelope Systems and Materials
P.O. Box 2008
Oak Ridge, TN 37831-6070
Phone: 423-574-4345
http://www.cad.ornl.gov/kch/demo.html

U.S. Department of Energy's EREN: Energy Efficiency and Renewable Energy Network
http://www.eren.doe.gov

U.S. Department of Energy
http://www.doe.gov

U.S. Government's Federal Information Network
http://www.fedworld.gov/

## Window Installation Guidelines

Residential Window and Door Installation Guide, Version 1.2
Association of Window and Door Installers
515 N. Flagler Drive
West Palm Beach, FL 33401
(407) 655-0696

CAWM 400-95 - Standard Practice for Installation of Windows with Integral Mounting Flange in Wood Frame Construction
California Association of Window Manufacturers
2080-A North Tustin Avenue
Santa Ana, California 92701-2148
(714) 835-2296

Window and Door Installation Certification Program Procedural Guide
National Certified Testing Laboratories
National Accreditation & Management Institute, Inc.
152 Leader Heights Rd.
York, PA 17403
(717) 741-9572

CSA A440.4 - Window and Door Installation
Canadian Standards Association
178 Rexdale Boulevard
Rexdale, Ontario M9W 1R3

CEGS Section 08520 - Aluminum Window Installation Guide Specification
Department of the Army
Engineer Division, Huntsville
CEHND-ED-ES (GS Section)
PO Box 1600
Huntsville, AL 35807-4301

Installing Wood Windows and Doors (videotape) and Participant's Notebook
National Wood Window and Door Association
1400 East Touhy Avenue, Suite G-54
Des Plaines, IL 60018
(847) 299-5200

# APPENDIX E

# Codes and Standards

## American Architectural Manufacturers Association (AAMA)

AAMA 502-90, "Voluntary Specifications for Sash Balances."

AAMA 910-93, "Voluntary Life Cycle Specifications and Test Methods for Architectural Grade Windows and Sliding Glass Doors."

AAMA 1503.1-1988, "Voluntary Test Method for Thermal Transmittance and Condensation Resistance of Windows, Doors and Glazed Wall Sections."

AAMA TIR-A10-1992, "Wind Loads on Components and Cladding for Buildings Less Than 90 Feet Tall."

AAMA CW #2-1979, "The Rain Screen Principle and Presssure Equalized Wall Design."

AAMA CW #11-1985, "Design Wind Loads for Buildings and Boundary Layer Wind Tunnel Testing."

AAMA/NWWDA, "Voluntary Specifications for Aluminum, Vinyl (PVC), and Wood Windows and Glass Doors–1996."

ANSI/AAMA 101-93, "Voluntary Specifications for Aluminum and Poly (Vinyl Chloride) (PVC) Prime Windows and Glass Doors."

ANSI/AAMA 1002.10-93, "Voluntary Specifications for Insulating Storm Products for Windows and Sliding Glass Doors."

## American Society of Heating, Refrigerating and Air Conditioning Engineers, Inc. (ASHRAE)

ANSI/ASHRAE Standard 74-1988, "Method of Measuring Solar-Optical Properties of Materials."

*ASHRAE Handbook of Fundamentals.*

## American Society for Testing and Materials (ASTM)

ASTM E 283-91, "Standard Test Method for Determining the Rate of Air Leakage Through Exterior Windows, Curtain Walls and Doors Under Specified Pressure Differences Across the Specimen."

ASTM E 330-90, "Standard Test Method for Structural Performance of Exterior Windows, Curtain Walls and Doors by Uniform Static Air Pressure Difference."

ASTM E 331-93, "Standard Test Method for Water Penetration of Exterior Windows, Curtain Walls and Doors by Uniform Static Air Pressure Difference."

ASTM E 413-94, "Classification for Rating Sound Insulation."

ASTM E 547-93, "Standard Test Method for Water Penetration of Exterior Windows, Curtain Walls and Doors by Cyclic Static Air Pressure Differential."

ASTM E 1332-90, "Classification for Determination of Outdoor-Indoor Transmission Class."

## Canadian Standards Association (CSA)

CAN/CSA-A440.1-M1990-M90. "User Selection Guide to CSA Standard CAN/CSA A440 M90."

CAN/CSA A440-M90/A440.1-M90. "Windows: User Selection Guide to CSA Standards."

CAN/CSA A440.2-93. "Energy Performance Evaluation of Windows and Sliding Glass Doors."

## Council of American Building Officials

*CABO Model Energy Code* 1995 Edition.

## International Standards Organization (ISO)

ISO 7345 (1987). "Thermal Insulation: Physical Quantities and Definitions."

ISO 9050 (1988). "Glass in Building: Determination of Light Transmittance, Direct Solar Transmittance, Total Solar Energy Transmittance, and Related Glazing Factors."

ISO 10293 (1992). "Glass in Building: Determination of Steady-State U Values (Thermal Transmittance) of Multiple Glazing: Heat Flow Meter Method."

ISO 10292 (1994). "Glass in Building: Calculation of Steady-State U Values (Thermal Transmittance) of Multiple Glazing."

ISO 10291 (1994). "Glass in Building: Determination of Steady State U Values (Thermal Transmittance) of Multiple Glazing: Guarded Hot Plate Method."

ISO 8990 (1994). "Thermal Insulation: Determination of Steady-State Thermal Transmission Properties: Calibrated and Guarded Hot Box."

ISO 12567 (1995). "Thermal Insulation: Determination of Thermal Resistance of Components: Hot Box Method For Doors and Windows."

## National Fenestration Rating Council (NFRC)

NFRC 100 (1997 edition). "Procedure for Determining Fenestration Product Thermal Properties (Currently Limited to U-values)."

NFRC 100-91. "Section B: Procedure for Determining Door System Product Thermal Properties (Currently Limited to U-values)."

NFRC 200-95. "Procedure for Determining Fenestration Product Solar Heat Gain Coefficients at Normal Incidence."

NFRC 300-94. "Procedures for Determining Solar Optical Properties of Simple Fenestration Products."

NFRC 301-93. "Standard Test Method for Emittance of Specular Surfaces Using Spectrometric Measurements."

NFRC 400-95. "Procedure for Determining Fenestration Product Air Leakage."

NFRC 900-95. "Procedure for Determining the Annual Heating and Cooling Energy Ratings of Fenestration Products Used in Residential Dwellings."

## National Wood Window and Door Association (NWWDA)

*Specifiers Guide to Wood Windows and Doors* [a compendium of NWWDA standards].

AAMA/NWWDA 101/I.S. 2-97, "Voluntary Specifications for Aluminum, Vinyl (PVC), and Wood Windows and Glass Doors."

NWWDA I.S.7-87, "Wood Skylight/Roof Windows."

NWWDA I.S.8-95, "Wood Swinging Patio Doors."

# Glossary

**AAMA.** American Architectural Manufacturers Association. A national trade association that establishes voluntary standards for the window, door, and skylight industry.

**Absorptance.** The ratio of radiant energy absorbed to total incident radiant energy in a glazing system.

**Acrylic.** A thermoplastic with good weather resistance, shatter resistance, and optical clarity, used for glazing.

**Aerogel.** A microporous, transparent silicate foam used as a glazing cavity fill material, offering possible U-values below 0.10 BTU/(h-sq ft-°F) or 0.56 W/(sq m-°C).

**Air infiltration.** The amount of air leaking in and out of a building through cracks in walls, windows, and doors.

**Air leakage rating.** A measure of the rate of infiltration around a window or skylight in the presence of a specific pressure difference. It is expressed in units of cubic feet per minute per square foot of window area (cfm/sq ft) or cubic feet per minute per foot of window perimeter length (cfm/ft). The lower a window's air leakage rating, the better its airtightness.

**Annealed glass.** Standard sheet of plate glass.

**Annealing.** Heating above the critical or recrystallization temperature, then controlled cooling of metal, glass, or other materials to eliminate the effects of cold-working, relieve internal stresses, or improve strength, ductility, or other properties.

**ANSI.** American National Standards Institute. Clearing house for all types of standards and specifications.

**Argon.** An inert, nontoxic gas used in insulating windows to reduce heat transfer.

**ASHRAE.** American Society of Heating, Refrigerating and Air Conditioning Engineers.

**ASTM.** American Society for Testing and Materials. Organization that sets standards for testing of materials.

**Awning.** Window similar to a casement except the sash is hinged at the top and always swings out.

**Balance.** A mechanical device (normally spring loaded) used in single- and double-hung windows as a means of counterbalancing the weight of the sash during opening and closing.

**Bay window.** An arrangement of three or more individual window units, attached so as to project from the building at various angles. In a three-unit bay, the center section is normally fixed, with the end panels operable as single-hung or casement windows.

**Bead.** A wood strip against which a swinging sash closes, as in a casement window. Also, a finishing trim at the sides and top of the frame to hold the sash, as in a fixed sash or a double-hung window. Also referred to as *bead stop*.

**Blackbody.** The ideal, perfect emitter and absorber of thermal radiation. It emits radiant energy at each wavelength at the maximum rate possible as a consequence of its temperature, and absorbs all incident radiance.

**BOCA.** Building Officials and Code Administrators.

**Bottom rail.** The bottom horizontal member of a window sash.

**Bow window.** A rounded bay window that projects from the wall in an arc shape, commonly consisting of five sashes.

**Brick molding.** A standard milled wood trim piece that covers the gap between the window frame and masonry.

**Btu (B.T.U.).** An abbreviation for British Thermal Unit—the heat required to increase the temperature of one pound of water one degree Fahrenheit.

**Casement.** A window sash that swings open on side hinges; in-swinging are French in origin; out-swinging are from England.

**Casing.** Exposed molding or framing around a window or door, on either the inside or outside, to cover the space between the window frame or jamb and the wall.

**Caulking.** A mastic compound for filling joints and sealing cracks to prevent leakage of water and air, commonly made of silicone, bituminous, acrylic, or rubber-based material.

**CFM.** Cubic Feet per Minute.

**Check rail.** The bottom horizontal member of the upper sash and the top horizontal member of the lower sash which meet at the middle of a double-hung window.

**Clerestory.** A window in the upper part of a lofty room that admits light to the center of the room.

**Composite frame.** A frame consisting of two or more materials—for example, an interior wood element with an exterior fiberglass element.

**Condensation.** The deposit of water vapor from the air on any cold surface whose temperature is below the dew point, such as a cold window glass or frame that is exposed to humid indoor air.

**Conduction.** Heat transfer through a solid material by contact of one molecule to the next. Heat flows from a higher-temperature area to a lower-temperature one.

**Convection.** A heat transfer process involving motion in a fluid (such as air) caused by the difference in density of the fluid and the action of gravity. Convection affects heat transfer from the glass surface to room air, and between two panes of glass.

**Crack length.** Total outside perimeter of window vent. Used in figuring air infiltration during AAMA certification testing.

**CRF.** Condensation Resistance Factor. An indication of a window's ability to resist condensation. The higher the CRF, the less likely condensation is to occur.

**Degree day.** A unit that represents a one-degree Fahrenheit deviation from some fixed reference point (usually 65° F) in the mean, daily outdoor temperature. *See also* heating degree day.

**Desiccant.** An extremely porous crystalline substance used to absorb moisture from within the sealed air space of an insulating glass unit.

**Dewpoint.** The temperature at which water vapor in air will condense at a given state of humidity and pressure.

**Divided light.** A window with a number of smaller panes of glass separated and held in place by muntins.

**DOE-2.1E.** A building-simulation computer program used to calculate total annual energy use.

**Double glazing.** In general, two thicknesses of glass separated by an air space within an opening to improve insulation against heat transfer and/or sound transmission. In factory-made double glazing units, the air between the glass sheets is thoroughly dried and the space is sealed airtight, eliminating possible condensation and providing superior insulating properties.

**Double-hung window.** A window consisting of two sashes of glass operating in a rectangular frame, in which both the upper and lower halves can be slid up and down. A counterbalance mechanism usually holds the sash in place.

**Double-strength glass.** Sheet glass between 0.115" and 0.133" (3–3.38 mm) thick.

**Drip.** A projecting fin or a groove at the outer edge of a sill, soffit, or other projecting member in a wall designed to interrupt the flow of water downward over the wall or inward across the soffit.

**Edge effects.** Two-dimensional heat transfer at the edge of a glazing unit due to the thermal properties of spacers and sealants.

**Electrochromics.** Glazing with optical properties that can be varied continuously from clear to dark with a low-voltage signal. Ions are reversibly injected or removed from an electrochromic material, causing the optical density to change.

**Electromagnetic spectrum.** Radiant energy over a broad range of wavelengths.

**Emergency exit window.** Fire escape window (egress window) large enough for a person to climb out. In U.S. building codes, each bedroom must be provided with an exit window. The exact width, area, and height from the floor are specified in the building codes.

**Emittance.** The ratio of the radiant flux emitted by a specimen to that emitted by a blackbody at the same temperature and under the same conditions.

**Evacuated glazing.** Insulating glazing composed of two glass layers, hermetically sealed at the edges, with a vacuum between to eliminate convection and conduction. A spacer system is needed to keep the panes from touching.

**Exterior stop.** The removable glazing bead that holds the glass or panel in place when it is on the exterior side of the light or panel, in contrast to an interior stop located on the interior side of the glass.

**Extrusion.** The process of producing vinyl or aluminum shapes by forcing heated material through an orifice in a die. Also, any item made by this process.

**Eyebrow windows.** Low, inward-opening windows with a bottom-hinged sash. These attic windows built into the top molding of the house are sometimes called "lie-on-your-stomach" or "slave" windows. Often found on Greek Revival and Italianate houses.

**Fanlight.** A half-circle window over a door or window, with radiating bars. Also called *circle top transom*.

**FCR.** Fenestration Cooling Rating. A rating number developed by the National Fenestration Rating Council to indicate relative performance during the cooling season. A higher FCR indicates better cooling season performance.

**Fenestration.** The placement of window openings in a building wall, one of the important elements in controlling the exterior appearance of a building. Also, a window or skylight and its associated interior or exterior elements, such as shades or blinds.

**FHR.** Fenestration Heating Rating. A rating number developed by the National Fenestration Rating Council to indicate relative performance during the heating season. A higher FHR indicates better heating season performance.

**Fiberglass.** A composite material made by embedding glass fibers in a polymer matrix. May be used as a diffusing material in sheet form, or as a standard sash and frame element.

**Fixed light.** A pane of glass installed directly into non-operating framing members; also, the opening or space for a pane of glass in a non-operating frame.

**Fixed panel.** An inoperable panel of a sliding glass door or slider window.

**Fixed window.** A window with no operating sashes.

**Flashing.** Sheet metal or other material applied to seal and protect the joints formed by different materials or surfaces.

**Float glass.** Glass formed by a process of floating the material on a bed of molten metal. It produces a high-optical-quality glass with parallel surfaces, without polishing and grinding.

**Fogging.** A deposit of contamination left on the inside surface of a sealed insulating glass unit due to extremes of temperatures or failed seals.

**Frame.** The fixed frame of a window which holds the sash or casement as well as hardware.

**Gas fill.** A gas other than air, usually argon or krypton, placed between window or skylight glazing panes to reduce the U-factor by suppressing conduction and convection.

**Glass.** An inorganic transparent material composed of silica (sand), soda (sodium carbonate), and lime (calcium carbonate) with small quantities of alumina, boric, or magnesia oxides.

**Glazing.** The glass or plastic panes in a window, door, or skylight.

**Glazing bead.** A molding or stop around the inside of a window frame to hold the glass in place.

**Greenhouse window.** A three-dimensional window that projects from the exterior wall and usually has glazing on all sides except the bottom, which serves as a shelf.

**Head track.** The track provided at the head of a sliding glass door. Also, the head member incorporating the track.

**Header.** The upper horizontal member of a window frame. Also called *head*.

**Heat-absorbing glass.** Window glass containing chemicals (with gray, bronze, or blue-green tint) which absorb light and heat radiation, and reduce glare and brightness. *See also* Tinted glass.

**Heat gain.** The transfer of heat from outside to inside by means of conduction, convection, and radiation through all surfaces of a house.

**Heating degree day.** Term used by heating and cooling engineers to relate the typical climate conditions of different areas to the amount of energy needed to heat and cool a building. The base temperature is 65 degrees Fahrenheit. A heating degree day is counted for each degree below 65 degrees reached by the average daily outside temperatures in the winter. For example, if on a given winter day, the daily average temperature outdoors is 30 degrees, then there are 35 degrees below the base temperature of 65 degrees. Thus, there are 35 heating degree days for that day.

**Heat loss.** The transfer of heat from inside to outside by means of conduction, convection, and radiation through all surfaces of a house.

**Heat-strengthened glass.** Glass that is reheated, after forming, to just below melting point, and then cooled, forming a compressed surface that increases its strength beyond that of typical annealed glass.

**Hinged windows.** Windows (casement, awning, and hopper) with an operating sash that has hinges on one side. *See also* Projected window.

**Hopper.** Window with sash hinged at the bottom.

**Horizontal slider.** A window with a movable panel that slides horizontally.

**Infiltration.** The movement of outdoor air into the interior of a building through cracks around windows and doors or in walls, roofs, and floors.

**Infrared radiation.** Invisible, electromagnetic radiation beyond red light on the spectrum, with wavelengths greater than 0.7 microns.

**Insulated shutters.** Insulating panels that cover a window opening to reduce heat loss.

**Insulating glass.** Two or more pieces of glass spaced apart and hermetically sealed to form a single glazed unit with one or more air spaces in between. Also called double glazing.

**Insulating value.** *See* U-factor.

**Insulation.** Construction materials used for protection from noise, heat, cold or fire.

**Interlocker.** An upright frame member of a panel in a sliding glass door which engages with a corresponding member in an adjacent panel when the door is closed. Also called *interlocking stile.*

**Jalousie.** Window made up of horizontally-mounted louvered glass slats that abut each other tightly when closed and rotate outward when cranked open.

**Jamb.** A vertical member at the side of a window frame, or the horizontal member at the top of the window frame, as in *head jamb.*

**Krypton.** An inert, nontoxic gas used in insulating windows to reduce heat transfer.

**KWH.** KiloWatt Hour. Unit of energy or work equal to one thousand watt-hours.

**Laminated glass.** Two or more sheets of glass with an inner layer of transparent plastic to which the glass adheres if broken. Used for safety glazing and sound reduction.

**Lift.** Handle for raising the lower sash in a double-hung window. Also called *sash lift.*

**Light.** A window; a pane of glass within a window. Double-hung windows are designated by the number of lights in upper and lower sash, as in six-over-six. Also spelled informally *lite.*

**Light-to-solar-gain ratio.** A measure of the ability of a glazing to provide light without excessive solar heat gain. It is the ratio between the visible transmittance of a glazing and its solar heat gain coefficient. Abbreviated LSG.

**Lintel.** A horizontal member above a window or door opening that supports the structure above.

**Liquid crystal glazing.** Glass in which the optical properties of a thin layer of liquid crystals are controlled by an electrical current, changing from a clear to a diffusing state.

**Long-wave infrared radiation.** Invisible radiation, beyond red light on the electromagnetic spectrum (above 3.5 micro meters), emitted by warm surfaces such as a body at room temperature radiating to a cold window surface.

**Low-conductance spacers.** An assembly of materials designed to reduce heat transfer at the edge of an insulating window. Spacers are placed between the panes of glass in a double- or triple-glazed window.

**Low-emittance (low-E) coating.** Microscopically thin, virtually invisible, metal or metallic oxide layers deposited on a window or skylight glazing surface primarily to reduce the U-factor by suppressing radiative heat flow. A typical type of low-E coating is transparent to the solar spectrum (visible light and short-wave infrared radiation) and reflective of long-wave infrared radiation.

**Meeting rail.** The part of a sliding glass door, a sliding window, or a hung window where two panels meet and create a weather barrier.

**Metal-clad windows.** Exterior wood parts covered with extruded aluminum or other metal, with a factory-applied finish to deter the elements.

**Micron.** One millionth ($10^{6}$) of a metric meter.

**Mil.** One thousandth of an inch, or 0.0254 millimeter.

**Mullion.** A major structural vertical or horizontal member between window units or sliding glass doors.

**Muntin.** A secondary framing member (horizontal, vertical, or diagonal) to hold the window panes in the sash. This term is often confused with mullion.

**Muntin grilles.** Wood, plastic, or metal grids designed for a single-light sash to give the appearance of muntins in a multilight sash, but removable for ease in cleaning the window.

**Nailing fin.** An integral extension of a window or patio door frame which generally laps over the conventional stud construction and through which nails are driven to secure the frame in place.

**NFRC.** National Fenestration Rating Council.

**Obscure glass.** Any textured glass (frosted, etched, fluted, ground, etc.) used for privacy, light diffusion, or decorative effects.

**Operable window.** Window that can be opened for ventilation.

**Operator.** Crank-operated device for opening and closing casement or jalousie windows.

**Pane.** One of the compartments of a door or window consisting of a single sheet of glass in a frame; also, a sheet of glass.

**Panel.** A major component of a sliding glass door, consisting of a light of glass in a frame installed within the main (or outer) frame of the door. A panel may be sliding or fixed.

**Panning.** In replacement window work, the outside aluminum trim that can extend around the perimeter of the window opening; used to cover up the old window material. Panning can be installed in the opening before the window, or can be attached directly to the window before installation.

**Particle dispersed glazing.** Glazing in which the orientation of small particles between two sheets of glass is controlled electrically, thus changing its optical properties.

**Parting stop.** A narrow strip, either integral or applied, that holds a sash or panel in position in a frame.

**Peak load.** The maximum thermal load to be provided by a heating or cooling system in a house.

**Photochromics.** Glazing with the optical properties that change in response to the amount of incident light.

**Picture window.** A large, fixed window framed so that it is usually, but not always, longer horizontally than vertically to provide a panoramic view.

**Pivot window.** A window with a sash that swings open or shut by revolving on pivots at either side of the sash or at top and bottom.

**Plastic film.** A thin plastic substrate, sometimes used as the inner layers in a triple- or quadruple-glazed window.

**Plastics.** Artificial substances made of organic polymers that can be extruded or molded into various shapes including window frames and sashes.

**Plate glass.** A rolled, ground, and polished product with true flat parallel plane surfaces affording excellent vision. It is now being replaced by float glass.

**Polyvinylchloride (PVC).** An extruded or molded plastic material used for window framing and as a thermal barrier for aluminum windows.

**Projected window.** A window fitted with one or more sashes opening on pivoted arms or hinges. Refers to casements, awnings, and hoppers.

**R-value.** A measure of the resistance of a glazing material or fenestration assembly to heat flow. It is the inverse of the U-factor ($R = 1/U$) and is expressed in units of hr-sq ft-°F/Btu. A high-R-value window has a greater resistance to heat flow and a higher insulating value than one with a low R-value.

**Radiation.** The transfer of heat in the form of electromagnetic waves from one separate surface to another. Energy from the sun reaches the earth by radiation, and a person's body can lose heat to a cold window or skylight surface in a similar way.

**Rail.** Horizontal member of a window sash.

**Reflectance.** The ratio of reflected radiant energy to incident radiant energy.

**Reflective glass.** Window glass coated to reflect radiation striking the surface of the glass.

**Refraction.** The deflection of a light ray from a straight path when it passes at an oblique angle from one medium (such as air) to another (such as glass).

**Relative humidity.** The percentage of moisture in the air in relationship to the amount of moisture the air could hold at that given temperature. At 100 percent relative humidity, moisture condenses and falls as rain.

**Retrofitting.** Adding or replacing items on existing buildings. Typical retrofit products are replacement doors and windows, insulation, storm windows, caulking, weatherstripping, vents, landscaping.

**RESFEN.** A computer program used to calculate energy use based on window selection in residential buildings.

**Roof window.** A fixed or operable window similar to a skylight placed in the sloping surface of a roof.

**Rough opening.** The opening in a wall into which a door or window is to be installed.

**Safety glass.** A strengthened or reinforced glass that is less subject to breakage or splintering.

**Sash.** The portion of a window that includes the glass and the framing sections directly attached to the glass, not to be confused with the complete frame into which the sash sections are fitted.

**Screen.** Woven mesh of metal, plastic, or fiberglass stretched over a window opening to permit air to pass through, but not insects.

**Sealant.** A compressible plastic material used to seal any opening or junction of two parts, such as between the glass and a metal sash, commonly made of silicone, butyl tape, or polysulfide.

**Shade screen.** A specially fabricated screen of sheet material with small narrow louvers formed in place to intercept solar radiation striking a window; the louvers are so small that only extremely small insects can pass through. Also called *sun screen*. Also, an awning with fixed louvers of metal or wood.

**Shading coefficient (SC).** A measure of the ability of a window or skylight to transmit solar heat, relative to that ability for 1/8-inch clear, double-strength, single glass. It is being phased out in favor of the solar heat gain coefficient, and is approximately equal to the SHGC multiplied by 1.15. It is expressed as a number without units between 0 and 1. The lower a window's solar heat gain coefficient or shading coefficient, the less solar heat it transmits, and the greater is its shading ability.

**Sheet glass.** A transparent, flat glass found in older windows, now largely replaced by float glass.

**Short-wave infrared radiation.** Invisible radiation, just beyond red light on the electromagnetic spectrum (between 0.7 and 2.5 microns), emitted by hot surfaces and included in solar radiation.

**Sill.** The lowest horizontal member in a door, window, or sash frame.

**Sill track.** The track provided at the sill of a sliding glass door. Also, the sill member incorporating such a track.

**Simulated divided lights.** A window that has the appearance of a number of smaller panes of glass separated by muntins, but actually is a larger glazing unit with the muntins placed between or on the surfaces of the glass layers.

**Single glazing.** Single thickness of glass in a window or door.

**Single-hung window.** A window consisting of two sashes of glass, the top one stationary and the bottom movable.

**Single-strength glass.** Glass with thickness between 0.085" and 0.100" (2.16–2.57 mm).

**Skylight (operable or pivot).** A roof window that gives light and ventilation.

**Sliding glass door.** A door fitted with one or more panels that move horizontally on a track and/or in grooves. Moving action is usually of rolling type (rather than sliding type). Also called *gliding door, rolling glass door,* and *patio sliding door.*

**Sliding window.** A window fitted with one or more sashes opening by sliding horizontally or vertically in grooves provided by frame members. Vertical sliders may be single- or double-hung.

**Smart window.** Generic term for windows with switchable coatings to control solar gain.

**Solar control coatings.** Thin film coatings on glass or plastic that absorb or reflect solar energy, thereby reducing solar gain.

**Solar heat gain coefficient (SHGC).** The fraction of solar radiation admitted through a window or skylight, both directly transmitted, and absorbed and subsequently released inward. The solar heat gain coefficient has replaced the shading coefficient as the standard indicator of a window's shading ability. It is expressed as a number between 0 and 1. The lower a window's solar heat gain coefficient, the less solar heat it transmits, and the greater its shading ability. SHGC can be expressed in terms of the glass alone or can refer to the entire window assembly.

**Solar radiation.** The total radiant energy from the sun, including ultraviolet and infrared wave lengths as well as visible light.

**Solar screen.** A sun shading device, such as screens, panels, louvers, or blinds, installed to intercept solar radiation.

**Solar spectrum.** The intensity variation of sunlight across its spectral range.

**Sound Transmission Class (STC).** The sound transmission loss rating of a material over a selected range of sound frequencies. The higher the number, the less sound transmitted.

**Spectrally selective glazing.** A coated or tinted glazing with optical properties that are transparent to some wavelengths of energy and reflective to others. Typical spectrally selective coatings are transparent to visible light and reflect short-wave and long-wave infrared radiation.

**Stile.** The upright or vertical edges of a door, window, or screen.

**Stool.** The shelf-like board of the interior part of the window sill, against which the bottom rail of the sash closes.

**Stop.** The molding on the inside of a window frame against which the window sash closes; in the case of a double-hung window, the sash slides against the stop. Also called *bead, side stop, window stop,* and *parting stop.*

**Storm windows.** A second set of windows installed on the outside or inside of the primary windows to provide additional insulation and wind protection.

**Sun control film.** A tinted or reflective film applied to the glazing surface to reduce visible, ultra-violet, or total transmission of solar radiation. Reduces solar heat gain in summer and glare. Some can be removed and reapplied with changing seasons.

**Superwindow.** A window with a very low U-factor, typically less than 0.15, achieved through the use of multiple glazings, low-E coatings, and gas fills.

**Switchable glazings.** Glazings with optical properties that can be reversibly switched from clear to dark or reflective.

**Tempered glass.** Treated glass that is strengthened by reheating it to just below the melting point and then suddenly cooling it. When shattered, it breaks into small pieces. Approximately five times stronger than standard annealed glass; is required as safety glazing in patio doors, entrance doors, side lights, and other hazardous locations. It cannot be recut after tempering.

**Thermal break.** An element of low conductance placed between elements of higher conductance to reduce the flow of heat. Often used in aluminum windows.

**Thermal expansion.** Change in dimension of a material as a result of temperature change.

**Thermal mass.** Mass in a building (furnishings or structure) that is used to absorb solar gain during the day and release the heat as the space cools in the evening.

**Thermochromics.** Glazing with optical properties that can change in response to temperature changes.

**Thermogram.** An image of an object taken with an infrared camera that shows surface temperature variations.

**Threshold.** The member that lies at the bottom of a sliding glass door or swinging door; the sill of a doorway.

**Tilt window.** A single- or double-hung window whose operable sash can be tilted into the room for interior washability.

**Tinted glass.** Glass colored by incorporation of a mineral admixture. Any tinting reduces both visual and radiant transmittance.

**Transmittance.** The percentage of radiation that can pass through glazing. Transmittance can be defined for different types of light or energy, e.g., visible light transmittance, UV transmittance, or total solar energy transmittance.

**Transom.** A horizontal transverse beam or bar in a frame; a crosspiece separating a door or the like from a window or fanlight above it. Also, a window above a door or other window, built on and commonly hinged to a transom.

**Transom window.** The window sash located above a door. Also called *transom light.*

**Triple glazing.** Three panes of glass or plastic with two air spaces between.

**U-factor (U-value).** A measure of the rate of non-solar heat loss or gain through a material or assembly. It is expressed in units of Btu/hr-sq ft-°F (W/sq m-°C). Values are normally given for NFRC/ASHRAE winter conditions of 0° F (18° C) outdoor temperature, 70° F (21° C) indoor temperature, 15 mph wind, and no solar load. The U-factor may be expressed for the glass alone or the entire window, which includes the effect of the frame and the spacer materials. The lower the U-factor, the greater a window's resistance to heat flow and the better its insulating value.

**UBC.** Uniform Building Code.

**Ultraviolet light (UV).** The invisible rays of the spectrum that are outside of the visible spectrum at its short-wavelength violet end. Ultraviolet rays are found in everyday sunlight and can cause fading of paint finishes, carpets, and fabrics.

**Vapor retarder.** A material that reduces the diffusion of water vapor across a building assembly.

**Vent.** The movable framework or sash in a glazed window that is hinged or pivoted to swing open.

**Vinyl.** Polyvinyl chloride material, which can be both rigid or flexible, used for window frames.

**Vinyl-clad window.** A window with exterior wood parts covered with extruded vinyl.

**Visible light.** The portion of the electromagnetic spectrum that produces light that can be seen. Wavelengths range from 380 to 720 nanometers.

**Visible transmittance (VT).** The percentage or fraction of the visible spectrum (380 to 720 nanometers) weighted by the sensitivity of the eye, that is transmitted through the glazing.

**Warm-edge technology.** The use of low-conductance spacers to reduce heat transfer near the edge of insulated glazing.

**Weatherstripping.** A strip of resilient material for covering the joint between the window sash and frame in order to reduce air leaks and prevent water from entering the structure.

**Weep hole.** A small opening in a wall or window sill member through which water may drain to the building exterior.

**Window.** A glazed opening in an external wall of a building; an entire unit consisting of a frame sash and glazing, and any operable elements.

**Window hardware.** Various devices and mechanisms for the window including catches, fasteners and locks, hinges, pivots, lifts and pulls, pulleys and sash weights, sash balances, and stays.

# References

## GENERAL REFERENCES

American Architectural Manufacturers Association. *Industry Statistical Review and Forecast.* Palatine, IL: American Architectural Manufacturers Association, 1995.

American Architectural Manufacturers Association. *Skylight Handbook: Design Guidelines.* Palatine, IL: American Architectural Manufacturers Association, 1988.

American Architectural Manufacturers Association. *Window Selection Guide.* Palatine, IL: American Architectural Manufacturers Association, 1988.

American Institute of Architects. "Daylighting Design." *Architect's Handbook of Energy Practice* (1982).

American Institute of Architects. "Shading and Sun Controls." *Architect's Handbook of Energy Practice* (1982).

American Society for Heating, Refrigerating and Air-Conditioning Engineers. "Energy-Efficient Window Design." *ASHRAE Technical Data Bulletin* 66, no. 2 (1990).

American Society of Heating, Refrigeration, and Air Conditioning Engineers. "Fenestration." In ASHRAE Handbook of Fundamentals. American Society of Heating, Refrigeration, and Air Conditioning Engineers, 1993.

Arasteh, D. "Advances in Window Technology: 1973-1993." In *Advances in Solar Energy, An Annual review of Research and Development 9.* Boulder, CO: American Solar Energy Society, Inc., 1994. *Lawrence Berkeley Laboratory Report 36891.*

Arasteh, D. "Super Windows." *Glass Magazine* 5 (1989): 82-83.

Canada Mortgage and Housing Corporation. *Door and Window Installation.* Ottawa, Ontario: Canada Mortgage and Housing Corporation, 1988.

Canada Mortgage and Housing Corporation. *Trouble Free Windows, Doors and Skylights.* Ottawa, Ontario: Canada Mortgage and Housing Corporation, 1988.

Collins, B. *Windows and People: A Literature Survey, Psychological Reaction with and without Windows.* National Bureau of Standards, Building Science Series 70, 1975.

Creative Homeowner Press. *Quick Guide: Windows & Doors.* Upper Saddle River, NJ: Creative Homeowner Press, 1994.

Fricke, J. "Aerogels." *Scientific American.* 258-5 (1988): 92-97.

Gilmore, V. E. "Superwindows." *Popular Science* 3 (1986): 76.

International Energy Agency Solar Heating and Cooling Programme. *Solar Energy Houses: Strategies, Technologies, Examples.* London, England: James & James Ltd., 1996.

International Energy Agency Solar Heating and Cooling Programme. *Passive Solar Commercial and Institutional Buildings: A Sourcebook of Examples and Design Insights.* West Sussex, England: John Wiley & Sons, Ltd., 1994.

Iqbal, M. *An Introduction to Solar Radiation.* Toronto, Ontario: Academic Press, 1983.

Kaufman, J., and H. Haynes, eds. *IES Lighting Handbook: Reference Volume.* Illuminating Engineering Society of North America, 1981.

Konzo, S. *Speaking of Windows.* Urbana, Champaign: University of Illinois Small Homes Council, 1984.

Lam, William, *Perception and Lighting as Formgivers for Architecture*. New York: McGraw-Hill Book Company, 1987.

Lawrence Berkeley Laboratory, "Windows for Energy Efficient Buildings," U.S. Department of Energy, Vol. 1, No. 2, January 1980, .

Mazria, E. *The Passive Solar Energy Book*. Emmaus, PA: Rodale Press, 1979.

McCluney, R. *Choosing the Best Window for Hot Climates*. Cape Canaveral, FL: Florida Solar Energy Center, 1993.

McCluney, R. *Introduction to Radiometry and Photometry*. Boston, MA: Artech House, 1994.

McCluney, R. *Fenestration Solar Gain Analysis*. Cape Canaveral, FL: Florida Solar Energy Center, 1996.

McCluney, R., and C. Gueymard. *Selecting Windows for South Florida Residences*. Cape Canaveral, FL: Florida Solar Energy Center, 1993.

McCluney, R., M. Huggins, and C. Emrich. *Fenestration Performance: An Annotated Bibliography*. Cape Canaveral, FL: Florida Solar Energy Center, 1990.

McGowan, A. *Energy-Efficient Residential and Commercial Windows Reference Guide*. Canadian Electricity Association, 1995.

National Research Council of Canada. "Window Performance and New Technology." In the proceedings of Building Science Insight Conference. Ottawa, Ontario: National Reserch Council of Canada, 1988.

Natural Resources Canada. *Consumer's Guide to Buying Energy-Efficient Windows and Doors*, Minister of Supply and Services, 1994.

Nisson, J. D. N. "Windows and Energy Efficiency: Principles, Practice and Available Products." *Energy Design Update* 31 (1985).

Ortho Books. *Doors, Windows & Skylights: Selecting and Installing*. San Ramon, CA: Ortho Books, 1992.

Passive Solar Industries Council. *Designing Low-Energy Buildings: Passive Solar Strategies and ENERGY-10 Software*. Washington, DC: Passive Solar Industries Council, 1996.

Selkowitz, S. "Smart Windows." *Glass Magazine* (Aug. 1986).

Selkowitz, S. and S. LaSourd. "Amazing Glazing." *Progressive Architecture* (June, 1994).

Schuman, J. "Cool Glazing." *Progressive Architecture* (April 1992).

Shurcliff, W. *Thermal Shutters and Shades: Over 100 Schemes for Reducing Heat-Loss Through Windows*. Andover, MA: Brick House Publishing, 1980.

Sunset Books. *Windows & Skylights*. Menlo Park, CA: Sunset Books, 1996.

Turner, Denis Philip, *Window Glass Design Guide*. London, England: The Architectural Press, 1977.

Warner, J. L. "Consumer Guide to Energy-Saving Windows." *Home Energy* 7, no. 4 (July/Aug 1990): 17-22.

Warner, J. L. "How to Avoid Window Condensation." *Home Energy* 8, no. 5 (Sept/Oct 1991): 27-29.

Warner, J., S. Reilly, S. Selkowitz, and D. Arasteh. "Utility and Economic Benefits of Electrochromic Smart Windows." In the proceedings of the ACEEE 1992 Summer Study on Energy Efficiency. Pacific Grove, CA: ACEEE, June 1992. *Lawrence Berkeley Laboratory Report 32638*.

## TECHNICAL REFERENCES

Arasteh, D. "Analysis of Frame and Edge Heat Transfer in Residential Windows." In the proceedings of Thermal Performance of the Exterior Envelopes of Buildings IV. Orlando, FL: Dec. 1989. *Lawrence Berkeley Laboratory Report 26068*.

Arasteh, D., F. Beck, B. Griffith, N. Byars, and M. Acevedo-Ruiz. "Using Infrared Thermography for the Study of Heat Transfer Through Building Envelope Components." *ASHRAE Transactions* 98, no. 1 (1992). *Lawrence Berkeley Laboratory Report 29752*.

Arasteh, D., F. Beck, N. Stone, W. duPont, and M. Koenig. "Phase I Results of the NFRC U-Value Procedure Validation Project." *ASHRAE Transactions* 100, no. 1 (1994). *Lawrence Berkeley Laboratory Report 34270*.

Arasteh, D., B. Griffith, and P. LaBerge. "Integrated Window Systems: An Advanced Energy-Efficient Residential Fenestration Product." In the proceedings of the 19th National Passive Solar Conference. San Jose, CA: American Solar Energy Society, Inc., 1994. *Lawrence Berkeley Laboratory Report 35417*.

Arasteh, D., J. Hartman, and M. Rubin. "Experimental Verification of a Model of Heat Transfer Through Windows," In the proceedings of the ASHRAE Winter Meeting, *Symposium on Fenestration Performance.* New York, NY: 1987.

Arasteh, D., R. Johnson, S. Selkowitz, and R. Sullivan. "Energy Performance and Savings Potentials with Skylights." *ASHRAE Transactions* 91, no. 1 (1984):154-179. *Lawrence Berkeley Laboratory Report 17457.*

Arasteh, D., R. Mathis, and W. duPont. "The NFRC Window U-Value Rating Procedure." In the proceedings of Thermal Performance of the Exterior Envelopes of Buildings V. Clearwater Beach, Florida: 1992. *Lawrence Berkeley Laboratory Report 32442.*

Arasteh, D., S. Reilly, and M. Rubin. "A Versatile Procedure for Calculating Heat Transfer Through Windows." Paper presented at ASHRAE Meeting. Vancouver, British Columbia: June 1989. *Lawrence Berkeley Laboratory Report 27534.*

Arasteh, D., and S. Selkowitz. "Prospects for Highly Insulating Window Systems." Paper presented at Conservation in Buildings: Northwest Perspective. Butte, MT: May 1985 (sponsored by the National Center for Appropriate Technology). *Lawrence Berkeley Laboratory Report 19492.*

Arasteh, D., and S. Selkowitz. "A Superwindow Field Demonstration Program in Northwest Montana." In the proceedings of Thermal Performance of the Exterior Envelopes of Buildings IV. Orlando, FL: Dec. 1989. *Lawrence Berkeley Laboratory Report 26069.*

Arasteh, D., S. Selkowitz, and J. Hartman. "Detailed Thermal Performance Data on Conventional and Highly Insulating Window Systems." In the proceedings of the BTECC Conference. Clearwater Beach, FL: 1985. *Lawrence Berkeley Laboratory Report 20348.*

Arasteh, D., S. Selkowitz, and J. Wolfe. "The Design and Testing of a Highly Insulating Glazing System for Use with Conventional Window Systems." *Journal of Solar Energy Engineering* 111 (1989): 44-53. *Lawrence Berkeley Laboratory Report 24903.*

Arasteh, D. et al. "Recent Technical Improvements to the WINDOW Computer Program." In the proceedings of the Windows Innovations Conference. Toronto, Ontario: CANMET, June 1995.

Arasteh, D. et. al. "WINDOW 4.1: A PC Program for Analyzing Window Thermal Performance in Accordance with Standard NFRC Procedures." Publ. LBL-35298, Lawrence Berkeley Laboratory, Energy & Environment Division, Berkeley, CA, 1993.

Arschehoug, O., M. Thyholt, I. Andresen, and B. Hugdal. "Frame and Edge Seal Technology: A State of the Art Survey." IEA Solar Heating and Cooling Program, Norwegian Institute of Technology. Trondheim, Norway: 1994.

Beck, F. "A Validation of the WINDOW4/FRAME3 Linear Interpolation Methodology." *ASHRAE Transactions* 100, no. 1 (1994). *Lawrence Berkeley Laboratory Report 34271.*

Beck, F., and D. Arasteh. "Improving the Thermal Performance of Vinyl-Framed Windows." In the proceedings of Thermal Performance of the Exterior Envelopes of Buildings V. Clearwater Beach, FL: 1992. *Lawrence Berkeley Laboratory Report 32782.*

Beck, F., B. Griffith, D. Turler, and D. Arasteh. "Using Infrared Thermography for the Creation of a Window Surface Temperature Database to Validate Computer Heat Transfer Models." In the proceedings of the Windows Innovations Conference. Toronto, Ontario: CANMET, June 1995. *Lawrence Berkeley Laboratory Report 36975.*

Bliss, R. W. "Atmospheric Radiation Near the Surface of the Ground." *Solar Energy* 5, no. 3:103 (1961).

Brambley, M. and S. Penner. "Fenestration Devices for Energy Conservation I: Energy Savings During the Cooling Season." Energy (Feb. 1979).

Brandle, K. and R. Boehm. "Air-Flow Windows: Performance and Applications." In the proceedings of Exterior Envelopes of Buildings Conference II : ASHRAE, Dec. 1982.

Burkhardt, W. C. "Solar Optical Properties of Gray and Brown Solar Control Series Transparent Acrylic Sheet." *ASHRAE Transactions* 81, no. 1 (1975): 384-97.

Burkhardt, W. C. "Acrylic Plastic Glazing: Properties, Characteristics and Engineering Data." *ASHRAE Transactions* 82, no. 1 (1976): 683.

Byars, N., and D. Arasteh. "Design Options for Low-Conductivity Window Frames." *Solar Energy Materials and Solar Cells* 25 (1992). Elsevier Science Publishers B.V. and *Lawrence Berkeley Laboratory Report 30498.*

Canadian Standards Association. *Windows.* Publication CAN/CSA-A440. Canadian Standards Association, 1990.

CANMET. *Long Term Performance of Operating Windows Subjected to Motion Cycling.* Ottawa, Ontario: CANMET. M91-7/235-1993E.

Carpenter, S., and A. McGowan. "Frame and Spacer Effects on Window U-Value." *ASHRAE Transactions* 95, no. 1, (1989).

Carpenter, S., and A. McGowan. "Effect of Framing Systems on the Thermal Performance of Windows." *ASHRAE Transactions* 99, no. 1, (1993).

Carpenter, S., and A. Elmahdy. "Thermal Performance of Complex Fenestration Systems." *ASHRAE Transactions* 100, no. 2 (1994).

Carpenter, S. and J. Hogan. "Recommended U-factors for Swinging, Overhead and Revolving Doors." *ASHRAE Transactions* (1996).

Collins, R., and S. Robinson. *Evacuated Glazing.* Sydney, Australia: University of Sydney Press, 1996.

Crooks, B. *Annual Energy Performance for Windows: Parametric Study.* Minneapolis, MN: Cardinal IG, 1994.

Crooks, B., J. Larsen, R. Sullivan, D. Arasteh, and S. Selkowitz. "NFRC Efforts to Develop a Residential Fenestration Annual Energy Rating Methodology." In the proceedings of Window Innovations Conference. Toronto, Ontario: 1995 and *Lawrence Berkeley Laboratory Report 36896.*

Curcija, D., and W. P. Goss. "Two-Dimensional Finite Element Model of Heat Transfer in Complete Fenestration Systems." *ASHRAE Transactions* 100, no. 2 (1994).

Curcija, D., and W. P. Goss. "Three-Dimensional Finite Element Model of Heat Transfer in Complete Fenestration Systems." In the proceedings of Window Innovations Conference. Toronto, Ontario: 1995.

Curcija, D., and W. P. Goss. "New Correlations for Convective Heat Transfer Coefficient on Indoor Fenestration Surfaces: Compilation of More Recent Work." In the proceedings of Thermal Performance of the Exterior Envelopes of Buildings VI. Clearwater Beach, Florida: 1995.

Curcija, D., W. P. Goss, J. P. Power, and Y. Zhao. *Variable-h' Model For Improved Prediction of Surface Temperatures in Fenestration Systems.* Amherst, MA: University of Massachusetts, 1996.

de Abreu, P., R. A. Fraser, H. F. Sullivan, and J. L. Wright. "A Study of Insulated Glazing Unit Surface Temperature Profiles Using Two-Dimensional Computer Simulation." *ASHRAE Transactions* 102, no. 2 (1996).

Duffie, J. A., and W. A. Beckman. *Solar Engineering of Thermal Processes.* New York: John Wiley & Sons, Inc., 1980.

Elmahdy, A. H. "A Universal Approach to Laboratory Assessment of the Condensation Potential of Windows." Paper presented at the 16th Annual Conference of the Solar Energy Society of Canada. Ottawa, Ontario: 1990.

Elmahdy, A. H., "Air Leakage Characteristics of Windows Subjected to Simultaneous Temperature and Pressure Differentials." In the proceedings of the Windows Innovations Conference. Toronto, Ontario: CANMET, 1995.

Elmahdy, A. H., and S. A. Yusuf. "Determination of Argon Concentration and Assessment of the Durability of High Performance Insulating Glass Units Filled with Argon Gas." *ASHRAE Transactions* (1995).

ElSherbiny, S. M., et al. "Heat Transfer by Natural Convection Across Vertical and Inclined Air Layers." *Journal of Heat Transfer* 104 (1982): 96-102.

Enermodal Engineering Ltd. "The Effect of Frame Design on Window Heat Loss: Phase 1." Enermodal Engineering Ltd. Ottawa, Ontario: 1987.

Ewing, W. B. and J. I. Yellott. "Energy Conservation through the Use of Exterior Shading of Fenestration. *ASHRAE Transactions* 82, no. 1 (1976):703-33.

Finlayson, E. U., et. al. *Window 4.0: Documentation of Calculation Procedures.* Berkeley, CA: Lawrence Berkeley Laboratory, 1993. *Lawrence Berkeley Laboratory Report 33943.*

Frost K., D. Arasteh, and J. Eto. *Savings from Energy Efficient Windows: Current and Future Savings from New Fenestration Technologies in the Residential Market.* Berkeley, CA: Lawrence Berkeley Laboratory, 1993. *Lawrence Berkeley Laboratory Report 33956.*

Galanis, N. and R. Chatiguy. "A Critical Review of the ASHRAE Solar Radiation Model." *ASHRAE Transactions* 92, no. 1 (1986).

Gates, D. M. "Spectral Distribution of Solar Radiation at the Earth's Surface." *Science* 151, no. 2 (1966): 3710.

Griffith, B., F. Beck, D. Arasteh, and D. Turler. "Issues Associated with the Use of Infrared Thermography for Experimental Testing of Insulated Systems." In the proceedings of Thermal Performance of the Exterior Envelopes of Buildings VI. Clearwater Beach, FL: December 1995. *Lawrence Berkeley Laboratory Report 36734.*

Griffith, B., D. Turler, and D. Arasteh. "Surface Temperature of Insulated Glazing Units: Infrared Thermography Laboratory Measurements." *ASHRAE Transactions* 102, no. 2 (1996).

Gueymard, C. A. "An Anisotropic Solar Irradiance Model for Tilted Surfaces and Its Comparison with Selected Engineering Algorithms." *Solar Energy* 38 (1987): 367-86.

Gueymard, C. A. "Development and Performance Assessment of a Clear Sky Spectral Radiation Model." In the proceedings of the 22nd Annual Solar Conference. Washington, DC: American Solar Energy Society, 1993.

Gueymard, C. A. "A Simple Model of the Atmospheric Radiative Transfer of Sunshine: Algorithms and Performance Assessment." Cape Canaveral, FL: Florida Solar Energy Center, 1995. *Florida Solar Energy Center Report FSEC-PF-270-95.*

Harrison, S., and S. van Wonderen. "A Test Method for the Determination of Window Solar Heat Gain Coefficient." *ASHRAE Transactions* 100, no. 1 (1994).

Hartman, J., M. Rubin, and D. Arasteh. "Thermal and Solar-Optical Properties of Silica Aerogel for Use in Insulated Windows." Paper presented at the 12th Annual Passive Solar Conference, Portland, OR: July 1987. *Lawrence Berkeley Laboratory Report 23386.*

Hickey, J. R., et al. "Observations of the Solar Constant and Its Variations: Emphasis on Nimbus 7 Results." Paper presented at the Symposium on the Solar Constant and the Special Distribution of Solar Irradiance. Hamburg, Germany: IAMAP 1981.

Hogan, J. F. "A summary of tested glazing U-values and the case for an industry wide testing program." *ASHRAE Transactions* 94, no. 2 (1988).

Illuminating Engineering Society of North America. *Recommended Practice of Daylighting.* New York: Illuminating Engineering Society of North America. 1979.

Johnson, B. *Heat Transfer through Windows.* Stockholm, Sweden: Swedish Council for Building Research. 1985.

Keyes, M. W. Analysis and Rating of Drapery Materials Used for Indoor Shading. *ASHRAE Transactions* 73, no. 1 (1967).

Klems, J. H. "Method of Measuring Nighttime U-Values Using the Mobile Window Thermal Test (MoWiTT) Facility." *ASHRAE Transactions* 98 no. 2 (1992). *Lawrence Berkeley Laboratory Report 30032.*

Klems, J. H. "Methods of Estimating Air Infiltration through Windows." *Energy and Buildings* 5 (1983): 243-252. *Lawrence Berkeley Laboratory Report 12891.*

Klems, J. H. "A New Method for Predicting the Solar Heat Gain of Complex Fenestration Systems: I. Overview and Derivation of the Matrix Layer Calculation." *ASHRAE Transactions* 100, no. 1 (1994a): 1065-1072.

Klems, J. H. "A New Method for Predicting the Solar Heat Gain of Complex Fenestration Systems: II. Detailed Description of the Matrix Layer Calculation." *ASHRAE Transactions* 100, no. 1 (1994a): 1073-1086.

Klems, J. H. and H. Keller. "Thermal Performance Measurements of Sealed Insulating Glass Units with Low-E Coatings Using the MoWitt Field Test Facility." Paper presented at the ASHRAE Winter Meeting, Symposium on Fenestration Performance. New York: Jan. 1987. *Lawrence Berkeley Laboratory Report 21583.*

Klems, J. H. and G. O. Kelley. "Calorimetric Measurements of Inward-Flowing Fraction for Complex Glazing and Shading Systems." *ASHRAE Transactions* (1995).

Klems, J. H. and J. L. Warner. "Measurement of Bidirectional Optical Properties of Complex Shading Devices." *ASHRAE Transactions* 101, no. 1 (1995): 791-801.

Klems, J. H., M. Yazdanian, and G. Kelley. "Measured Performance of Selective Glazings." In the proceedings of Thermal Performance of the Exterior Envelopes of Buildings VI. Clearwater Beach, FL: 1995. *Lawrence Berkeley Laboratory Report 37747.*

Lampert, C. "Chromogenic Switchable Glazing: Towards the Development of the Smart Window." In the proceedings of Window Innovations Conference. Toronto, Canada: CANMET, 1995. *Lawrence Berkeley Laboratory Report 37766.*

Lawrence Berkeley National Laboratory. "WINDOW 4.0: Program Description." Berkeley, CA: Lawrence Berkeley National Laboratory, 1992. *Lawrence Berkeley Laboratory Report 32091.*

Lee, E., D. Hopkins, M. Rubin, D. Arasteh, and S. Selkowitz. "Spectrally Selective Glazings for Residential Retrofits in Cooling-Dominated Climates." *ASHRAE Transactions* 100, no. 1 (1994). *Lawrence Berkeley Laboratory Report 34455.*

Mathis, R. C. and R. Garries. Instant, annual life: A discussion on the current practice and evolution of fenestration energy performance rating." In the proceedings of Window Innovations Conference. Toronto, Canada: CANMET, 1995.

McCluney, R. "Determining Solar Radiant Heat Gain of Fenestration Systems." *Passive Solar Journal* 4, no. 4 (1987):439-87.

McCluney, R. "The Death of the Shading Coefficient?" ASHRAE Journal (March 1991): 36-45.

McCluney, R. "Sensitivity of Optical Properties and Solar Gain of Spectrally Selective Glazing Systems to Changes in Solar Spectrum" In the proceedings of the 22nd Annual Solar Conference. Washington, DC: American Solar Energy Society, 1993.

McCluney, R. "Angle of Incidence and Diffuse Radiation Influences on Glazing System Solar Gain," In the proceedings of the Annual Solar Conference. San Jose, CA: American Solar Energy Society, 1994.

McCluney, R. "Sensitivity of Fenestration Solar Gain to Source Spectrum and Angle of Incidence." *ASHRAE Transactions* 10 (June 1996).

McCluney, R., and L. Mills. "Effect of Interior Shade on Window Solar Gain." *ASHRAE Transactions* 99, no. 2 (1993).

Moore, G. L. and C. W. Pennington. "Measurement and Application of Solar Properties of Drapery Shading Materials." *ASHRAE Transactions* 73, no. 1 (1967).

Ozisik, N. and L. F. Schutrum. "Solar Heat Gain Factors for Windows with Drapes." *ASHRAE Transactions* (1960).

Papamichael, K. "New Tools for the Analysis and Design of Building Envelopes." In the proceedings of Thermal Performance of the Exterior Envelopes of Buildings VI. Clearwater Beach, FL: 1995. *Lawrence Berkeley Laboratory Report 36281.*

Parmelee, G. V. and R. G. Huebscher. "Forced Convection Heat Transfer from Flat Surfaces." *ASHVE Transactions* (1947): 245-84.

Parmelee, G. V. and W. W. Aubele. "Radiant Energy Emission of Atmosphere and Ground." *ASHRAE Transactions* (1952).

Patenaude, A. "Air Infiltration Rate of Windows Under Temperature and Pressure Differentials." In the proceedings of Window Innovation Conference. Toronto, Ontario: CANMET, 1995.

Pennington, C. W. "How Louvered Sun Screens Cut Cooling, Heating Loads." *Heating, Piping, and Air Conditioning* (Dec. 1968).

Pennington, C. W., et al. "Experimental Analysis of Solar Heat Gain through Insulating Glass with Indoor Shading." *ASHRAE Journal* 2 (1964).

Pennington, C. W. and G. L. Moore. "Measurement and Application of Solar Properties of Drapery Shading Materials." *ASHRAE Transactions* 73, no. 1 (1967).

Pennington, C. W., C. Morrison, and R. Pena. "Effect of Inner Surface Air Velocity and Temperatures Upon Heat Loss and Gain through Insulating Glass." *ASHRAE Transactions* 79, no. 2 (1973).

Perez, R., et al. "An Anisotropic Hourly Diffuse Radiation Model for Sloping Surfaces–Description, Performance Validation, and Site Dependency Evaluation." Solar Energy 36 (1986): 481-98.

Reilly, M. S., D. Arasteh, and S. Selkowitz. "Thermal and Optical Analysis of Switchable Window Glazings." *Solar Energy Materials* 22 (1991). Elsevier Science Publishers B.V. *Lawrence Berkeley Laboratory Report 29629.*

Rubin, M. "Calculating Heat Transfer Through Windows." *International Journal of Energy Research* 6, no. 4 (1982): 341-349. TA-84, LBL-12486.

Rubin, M. "Optical Constants and Bulk Optical Properties of Soda Lime Silica Glasses for Windows."

Berkeley, CA: Lawrence Berkeley Laboratory, 1984. *Lawrence Berkeley Laboratory Report 13572.*

Rubin, M. "Optical Properties of Soda Lime Silica Glasses." *Solar Energy Materials* 12 (1985).

Rubin, M. "Solar Optical Properties of Windows." *International Journal of Energy Research* 6 (1982):123-133. *Lawrence Berkeley Laboratory Report 12246.*

Rudoy, W. and F. Duran. "Effect of Building Envelope Parameters on Annual Heating/Cooling Load." *ASHRAE Journal* 7 (1975).

Selkowitz, S. "Influence of Windows on Building Energy Use." Paper presented at the Windows in Building Design and Maintenance, Gothenburg, Sweden, June 1984.

Selkowitz, S., and Bazjanac, V. "Thermal Performance of Managed Window Systems." *ASHRAE Transactions* 85, no. 1 (1981): 392-408. *Lawrence Berkeley Laboratory Report 09933.*

Selkowitz, S. "Thermal Performance of Insulating Window Systems." *ASHRAE Transactions* 85, no. 2 (1981). *Lawrence Berkeley Laboratory Report 08835.*

Selkowitz, S. "Window Performance and Building Energy Use: Some Technical Options for Increasing Energy Efficiency." In the proceedings of Energy Sources: Conservation and Renewables, AIP Conference. Washington, DC: April 1985. *Lawrence Berkeley Laboratory Report 20213.*

Selkowitz, S., M. Rubin, E. Lee, and R. Sullivan. "A Review of Electrochromic Window Performance Factors." In the proceedings of SPIE International Symposium on Optical Materials Technology for Energy Efficiency and Solar Energy Conversion XIII. Friedrichsbau, Freiburg, Germany: April 1994. *Lawrence Berkeley Laboratory Report 35486.*

Selkowitz, S., and C. Lampert. "Application of Large-Area Chromogenics to Architectural Glazings." In *Large Area Chromogenics: Materials and Devices for Transmittance Control,* edited by C. Lampert. SPIE, 1989. *Lawrence Berkeley Laboratory Report 28012.*

Selkowitz, S. and R. Sullivan. "Analysis of Window Performance in a Single-Family Residence." In the proceedings of 9th National Passive Solar Conference, eds. J. Hayes and A. Wilson. International Solar Energy Society, 1984. *Lawrence Berkeley Laboratory Report 18247.*

Smith, W. A. and C. W. Pennington. "Shading Coefficients for Glass Block Panels." *ASHRAE Journal* 5 (Dec. 1964).

Sodergren, D. and T. Bostrom. "Ventilating with the Exhaust Air Window." *ASHRAE Journal* 13, no. 4 (1971).

Sterling, E. M. A., Arundel, T. D. Sterling. "Criteria for Human Exposure in Occupied Buildings." *ASHRAE Transactions* 91, no. 1 (1985).

Sullivan, H. F., J. L. Wright, and R. A. Fraser. "Overview of a Project to Determine the Surface Temperatures of Insulated Glazing Units: Thermographic Measurement and 2-D Simulation." *ASHRAE Transactions* 102, no. 2 (1996).

Sullivan, R., F. Beck, D. Arasteh, and S. Selkowitz. "Energy Performance of Evacuated Glazings in Residential Buildings." Paper presented at ASHRAE Conference. San Antonio, TX: June 1996. *Lawrence Berkeley Laboratory Report 37130.*

Sullivan, R., B. Chin, D. Arasteh, and S. Selkowitz. "RESFEN: A Residential Fenestration Performance Design Tool." *ASHRAE Transactions* 98, no. 1 (1992). *Lawrence Berkeley Laboratory Report 31176.*

Sullivan, R., K. Frost, D. Arasteh, and S. Selkowitz. "Window U-Value Effects on Residential Cooling Load." Berkeley, CA: Lawrence Berkeley National Laboratory, 1993. *Lawrence Berkeley Laboratory Report 34648.*

Sullivan, R., E. Lee, K. Papamichael, M. Rubin, and S. Selkowitz. "Effect of Switching Control Strategies on the Energy Performance of Electrochromic Windows." In the proceedings of SPIE International Symposium on Optical Materials Technology for Energy Efficiency and Solar Energy Conversion XIII. Friedrichsbau, Freiburg, Germany: April 1994. *Lawrence Berkeley Laboratory Report 35453.*

Sullivan, R., M. Rubin, and S. Selkowitz. "Reducing Residential Cooling Requirements Through the Use of Electrochromic Windows." In the proceedings of Thermal Performance of the Exterior Envelopes of Buildings VI. Clearwater Beach, FL: Dec. 1995. *Lawrence Berkeley Laboratory Report 37211.*

Sullivan, R and S. Selkowitz. "Energy Performance Analysis of Fenestration in a Single-Family Residence." *ASHRAE Transactions* 91, no. 2 (1984). *Lawrence Berkeley Laboratory Report 18561.*

Sullivan, R. and S. Selkowitz. "Residential Window Performance Analysis Using Regression Procedures." In the proceedings of CLIMA 2000 World Conference on Heating, Ventilation, and Air-Conditioning. Copenhagen, Denmark: August 1985. *Lawrence Berkeley Laboratory Report 19245.*

Sullivan, R, and S. Selkowitz. "Window Performance Analysis in a Single-Family Residence." In the proceedings of the BTECC Conference. Clearwater Beach, FL: 1985. *Lawrence Berkeley Laboratory Report 20079.*

Sullivan, R., and S. Selkowitz. "Residential Heating and Cooling Energy Cost Implications Associated with Window Types." *ASHRAE Transactions* 93, no. 1 (1986): 1525-1539. *Lawrence Berkeley Laboratory Report 21578.*

Sullivan, R. and S. Selkowitz. "Fenestration Performance Analysis Using an Interactive Graphics-Based Methodology on a Microcomputer." *ASHRAE Transactions* 95, no. 1 (1989). *Lawrence Berkeley Laboratory Report 26070.*

Sweitzer, G., D. Arasteh, and S. Selkowitz. "Effects of Low-E Glazing on Energy Use Patterns in Non-residential Daylighting Buildings." *ASHRAE Transactions* 93, no. 1 (1986): 1553-1566. *Lawrence Berkeley Laboratory Report 21577.*

Terman, M., S. Fairhurst, B. Perlman, J. Levitt, and R. McCluney. "Daylight Deprivation and Replenishment: A Psychobiological Problem with a Naturalistic Solution." In the proceedings of International Daylighting Conference. Long Beach, CA: 1986.

Van Dyke, R.L. and T.P. Konen. *Energy Conservation through Interior Shading of Windows: An Analysis, Test and Evaluation of Reflective Venetian Blinds.* Lawrence Berkeley Laboratory, March 1982. *Lawrence Berkeley Laboratory Report 14369,*

Vild, D. J. "Solar Heat Gain Factors and Shading Coefficients." *ASHRAE Journal* 10 (1964):47.

Weidt, J., and S. Selkowitz. "Field Air Leakage of Newly Installed Residential Windows." *ASHRAE Transactions* 85, no. 1 (1981):149-159. *Lawrence Berkeley Laboratory Report 09937.*

Wright, J. L. "Summary and Comparison of Methods to Calculate Solar Heat Gain," *ASHRAE Transactions* 101, no. 1 (1995).

Wright, J. L. "A Correlation to Quantify Convective Heat Transfer between Window Glazings." *ASHRAE Transactions* (1996).

Wright, J. L., R. Fraser, P. de Abreu, and H. F. Sullivan. "Heat Transfer in Glazing System Edge-Seals: Calculations Regarding Various Design Options." *ASHRAE Transactions* 100, no. 1 (1994).

Wright, J. L., and H. F. Sullivan. "A 2-D numerical model for natural convection in a vertical, rectangular window cavity." *ASHRAE Transactions* 100, no. 2 (1995).

Wright, J. L., and H. F. Sullivan. "A 2-D numerical model for glazing system thermal analysis." *ASHRAE Transactions* 100, no. 1 (1995).

Wright, J. L., and H. F. Sullivan. 1995. "A Simplified Method for the Numerical Condensation Resistance Analysis of Windows." In the proceedings of Window Innovations Conference. Toronto, Ontario: June 1995.

Yazdanian, M., and J. Klems. "Measurement of the Exterior Convective Film Coefficient for Windows in Low-Rise Buildings." *ASHRAE Transactions* 100, no. 1 (1994). *Lawrence Berkeley Laboratory Report 34717.*

Yellott, J. I. "Selective reflectance: A New Approach to Solar Heat Control." ASHRAE Transactions 69 (1963): 418.

Yellott, J. I. "Drapery Fabrics and Their Effectiveness in Sun Control." ASHRAE Transactions 71, no. 1 (1965): 260-72.

Yellott, J. I. "Shading Coefficients and Sun-Control Capability of Single Glazing." ASHRAE Transactions 72, no. 1 (1966): 72.

Yellott, J. I. "Effect of Louvered Sun Screens upon Fenestration Heat Loss." ASHRAE Transactions 78, no. 1 (1972): 199-204.

# Index

## A

AAMA, *see* American Architectural Manufacturers Association (AAMA)
absorptance, 43, 47–48
acrylic, 50, 62
aerogel, 82
air leakage, *see* infiltration
American Architectural Manufacturers Association (AAMA), 102, 152, 156
ASTM, 156
ASTM E-283, 40

## B

blinds
    heat gain reduced with, 130
    light diffused with, 117, 118–19
    privacy with, 114
building codes and standards, 120, 171–72

## C

casement windows, *see* hinged windows
center-of-glass U-factor, 29
coatings
    pyrolytic, 59, 68, 69
    solar control, 17
    sputtered, 68–69
    visible transmittance through, 38, 56
    *see also* low-emittance coatings; reflective coatings; spectrally selective coatings
comfort, thermal, 150–51
condensation, 151–52
    with aluminum frames, 99, 151–52
    between glazings, 153–55
    on NFRC label, 23, 152
    outdoor, 153
    prevention in installation, 107
Condensation Resistance Factor (CRF), 152
conduction, 24
convection, 24, 26
cool windows, 20
Cooling Rating (CR), 160, 163
    on NFRC label, 23
CR, *see* Cooling Rating (CR)
CRF, 152
curtains, *see* drapes

## D

daylight, 111, 114–15
    balanced lighting, 115–17
    diffusing direct, 115, 117, 118–19
    glare from, 117–18, 149
    outdoor surfaces reflecting, 117
    selecting window, amount of in, 147–48
    from sky, 119–20
DOE 2.1E, 163
double-glazed windows, 15, 16
    coatings and tints in, 59
    condensation between, 153–55
    construction of, 57–59
    divided lights in, 61–62
    energy performance with, 52, 59
    gap width in, 60
    insulating value of, 26
    *see also* insulating glass units (IGUs)
double-hung windows, *see* sliding windows
drapes
    heat gain reduced with, 129–30
    heat loss reduced with, 134
    light diffused with, 117, 118
    privacy with, 114
durability, 156

# E

edge effects, 29, 74
electrochromic glazing, 84–86
emittance, 43, 48
energy efficiency
  national impact of, 17–20
  technological advances in, 11–13, 16–20
energy performance
  with air leakage controlled, 38–40
  building codes for, 171–72
  of coatings, 48, 66–68, 77
  determining, 7–8, 158–68
  with double glazing, 59
  of insulating value, 25–31
  maintaining, 171
  peak loads reduced, 168–71
  properties of windows, 158–59
  rating system for, 22–23, 158–63
  of SHGC, 36
  with solar radiation controlled, 32–36
  of superwindows, 75–76, 79
  of vinyl frames, 79–80
Energy Star, 160
evacuated windows, 81
exterior shading devices
  heat gain increased with, 138
  heat gain reduced with, 33, 127–29
  light diffused with, 118

# F

fading, 44, 149–50
fiber-reinforced plastic, 50
fixed windows, 92
folding patio doors, 89
frames
  aluminum, 96–97, 99–101
  appearance of, 144–45
  condensation on, 151–52, 153
  engineered thermoplastic, 103–4
  fiberglass, 103–4
  heat loss from, 17, 29–30
  hybrid and composite, 104
  materials for, 96–104
  vinyl, 79–80, 97, 101–3
  wood, 96, 97–99
  wood composite, 104
French doors, 14, 89, 94

# G

garden windows, 91
gas fills, 16, 60, 70–71
glare control, 117–18, 149
glass
  float, 15, 49
  history of, 13–15
  plastic compared to, 51
  plate, 14–15
  properties of, 49–50
  tinted, see tinted glass
  U-factor of, 29
glass blocks, 62, 119
glazing materials, 41–42
  appearance of, 145–46
  properties of, in energy transfer, 42–48
  see also specific materials
greenhouse windows, 91

# H

heat gain, decreasing, 111, 158
  with coatings, 34, 65
  with double glazing, 59
  with exterior shading, 127–29
  glazing area reduction in, 124–26
  with insulating value, 28
  with interior shading devices, 129–31
  with landscaping, 129
  with overhangs, 127–29
  with tinted glass, 33
  by window orientation, 123–24
heat gain, determining, 34–36
heat gain, increasing, 32–33, 111
  coatings for, 33, 44–45
  with passive solar design, 136–38, 140–42
  solar access for, 138
  windows oriented for, 138–40
heat loss, decreasing, 111, 157
  glazing amount in, 131–34
  with insulating value, 27
  with shades and shutters, 134–36
  winds, by minimizing, 136
heat-strengthened glass, 49
Heating Rating (HR), 160, 163
  on NFRC label, 23
hinged windows, 89
  air leakage of, 38, 92–93
  ventilation with, 122, 155

house design
  climate and site sensitivity in, 10–11
  window technology effect on, 110–42
HR, *see* Heating Rating (HR)

# I

IGUs, *see* insulating glass units (IGUs)
infiltration, 25
  climate effect on, 39
  condensation, with, 153
  defined, 38, 159
  installation to prevent, 107
  measuring, 40, 95
  on NFRC label, 23, 158
  reducing, 38
  wind effect on, 136
infrared thermography, 77–80
installation
  air and vapor retarders in, 107
  of replacement windows, 108–9
  of skylights and roof windows, 107–8
  wall framing for, 105–6
insulating glass units (IGUs), 15
  energy comparison of, 77–79
  heat gain reduced with, 28
  heat loss reduced with, 27
  sealants for, 72–74
  *see also* double-glazed windows
insulating value, 24, 25–28
  defined, 25
  determining, 28–31
  energy effects of, 31
  heat gain reduced with, 28
  heat loss reduced with, 27
interior shading devices
  cost of, 173
  heat gain increased with, 138
  heat gain reduced with, 129–30
  light diffused with, 117, 118–19
  privacy with, 114

# J

jalousie windows, 91

# L

laminated glass, 50, 52, 70
landscaping
  glare reduced with, 118
  heat gain increased with, 138
  heat gain reduced with, 33, 129
Lawrence Berkeley National Laboratory, 18
Light-to-Solar-Gain (LSG) ratio, 38, 54–55
liquid crystal glazing, 84
low-conductance gas fills, *see* gas fills
low-E coatings, *see* low-emittance coatings
low-emittance coatings, 16
  detection of, 46
  in double glazing, 64–70
  energy performance of, 48, 66–68, 77
  heat loss reduced with, 44–45, 52
  placement of, 68
  types of, 66, 68–70
LSG, *see* Light-to-Solar-Gain (LSG) ratio

# M

maintenance, 156, 173–74
Model Energy Code for the United States, 172
multiple-pane units, 15, 16, 63–64
  energy impact of, 59
  gap width in, 60
muntins, 61

# N

National Fenestration Rating Council (NFRC)
  energy rating system by, 22–23, 158, 159–60
  long-term energy performance ratings by, 171
  overall U-factor calculations by, 30
  purposes of, 5, 172
  whole window numbers, 7
NFRC label, 7
  air leakage rating on, 23, 158
  availability of, 178
  condensation resistance rating, 23, 152
  Cooling Rating (CR) on, 23
  Heating Rating (HR) on, 23
  solar heat gain coefficient (SHGC) on, 23, 148,
    158
  U-factor on, 23, 158
  visible light transmittance (VT) on, 23, 37, 147,
    148, 158

## O

OITC, *see* Outdoor-Indoor Transmission Class
    (OITC) rating
one-way glass, 47
Outdoor-Indoor Transmission Class (OITC)
    rating, 156
overhangs
    glare reduction with, 118
    heat gain increased with, 138
    heat gain reduced with, 33, 127–29

## P

panning frame, 108
particle dispersed glazing, 84
passive solar heating, 11, 136–37, 140–42
patio doors, folding, 89
peak heating and cooling loads, 168–71
photochromics, 83
plastic glazing, 50–51, 62
plate glass, 14–15
polycarbonate, 50, 62
polyester, 51
pouring and debridging, 100
privacy, 114
projected windows, *see* hinged windows

## Q

quadruple-glazed windows, 63

## R

radiation, 24, 26–27
reflectance, 43, 45–47
reflective coatings, 47, 56–57
    detection of, 46
    in double glazing, 59
    heat gain reduced with, 34, 52
RESidential FENestration program (RESFEN)
    uses for, 163, 166, 168, 176
    using, 163
    Web availability of, 8, 163, 190
roof windows, 90
    condensation on, 153
    daylight through, 119
    insulating, 107–8
    thermal performance of, 27
    ventilation with, 90, 123
R-value, 28

## S

sashes
    heat loss from, 17, 29–30
    operating types, 88–92
    performance of, 92–94
    replacement, 108–9
SC, *see* shading coefficient (SC)
sealant, 57, 72, 74
shades
    heat gain increase with, 138
    heat gain reduced with, 130–31
    heat loss reduced with, 134
    light diffused with, 117, 118
shading, *see* exterior shading devices; interior
    shading devices
shading coefficient (SC), 25, 34, 35–36
SHGC, *see* solar heat gain coefficient (SHGC)
shutters, insulated, 135–36
single-glazed windows, 15
    condensation on, 151
    insulating value of, 25–26
skylights, 90
    condensation on, 153
    daylight through, 119–20
    double-glazed, 62
    insulation of, 107
    plastic glazing for, 50
    thermal performance of, 27
    ventilation with, 90, 123
sliding glass doors, 89, 94
sliding windows, 89
    air leakage of, 93
    ventilation with, 122–23, 155
smart windows, 20–21, 83–86
solar control coatings, 17
solar heat gain coefficient (SHGC), 25, 34, 159
    conversion from SC, 36
    energy effect of, 36
    on NFRC label, 23, 148, 158
solariums, 91
solar radiation
    determining heat gain from, 34–36
    energy transfer through glazing materials, 42–48
    heat gain from, 25, 32–33
    minimizing, 33–34
solar screen, 128
sound control, 155–56
Sound Transmission Class (STC) rating, 155–56
spacers
    double-seal, 72–73
    energy performance of, 77

functions of, 57, 72
materials for, 72, 73–74
single-seal, 72
warm edge, 16, 74
spectrally selective coatings
heat loss reduced with, 33, 34, 44–45
use of, 52, 64, 69
STC, see Sound Transmission Class (STC) rating
storm windows, 27
sunlight, see daylight
sun rooms, 91
superwindows, 20, 75
condensation on, 151
energy performance of, 75–76, 79
heat loss decreased with, 132–34

**T**

tempered glass, 49–50
thermal break, 99–100
thermochromics, 83–84
thermography, 77–80
tinted glass
colors of, 52
in double glazing, 59
Light-to-Solar-Gain (LSG) ratio of, 54–55
privacy with, 114
reasons for, 52
solar radiation minimized with, 33
spectrally selective, 38, 53–54
traditional, 53
transparency of, 42
uses for, 55–56
transmittance, 43–45
trickle ventilators, 121
triple-glazed windows, 15, 16
construction of, 63–64
energy impact of, 59

**U**

U-factor, 24
center-of-glass, 29
defined, 28, 159
edge effects in determining, 29, 74
energy effects of, 31
of frames, 29–30
of glass, 29
on NFRC label, 23, 158
overall, 30–31
of sashes, 29–30

see also insulating value
ultraviolet (UV) light, 32, 42
fading with, 44, 149
through glass, 43
U.S. Department of Energy, 5, 160
U-value, see U-factor

**V**

ventilation, 111, 155
building codes for, 120
cross-ventilation, 122–23
energy efficiency of, 121
trickle ventilators, 121
view, 111, 112–14
visible light, 32, 42, 149
visible light transmittance (VT), 25, 36–38, 44
defined, 36, 159
on NFRC label, 23, 37, 147, 148, 158
of reflective coatings, 38, 56
of tinted glazings, 38, 54

**W**

wall framing, 105–6
water-white glass, 36, 49
weatherstripping
infiltration stopped with, 17, 38
leakage of, 95
types of, 94–95
whole product performance, 22
wind, 136
windows
appearance of, 144–46
assembly of, 87–109
combining, 91–92
costs of, 172–77
emergency egress through, 95
energy performance of, 24–40, 157–72
function of, 146–56
glazing materials for, 41–86
history of, 13–15
in house design, 110–42
installation of, 105–9
rating system for, 22–23
replacement, 108–9
security of, 95–96, 113–14
selection of, 21, 143–79
World Wide Web, 8